The BEST WRITING on MATHEMATICS

2018

Mircea Pitici, Editor

PRINCETON UNIVERSITY PRESS
PRINCETON AND OXFORD

Published by Princeton University Press
41 William Street, Princeton, New Jersey 08540
6 Oxford Street, Woodstock, Oxfordshire OX20 1TR

press.princeton.edu

All Rights Reserved
ISBN 978-0-691-18276-6

Editorial: Vickie Kearn and Lauren Bucca
Production Editorial: Nathan Carr
Production: Jacquie Poirier
Publicity: Sara Henning-Stout and Kathryn Stevens
Copyeditor: Paula Bérard

This book has been composed in Perpetua.

Printed on acid-free paper. ∞

Printed in the United States of America

1 3 5 7 9 10 8 6 4 2

The **BEST WRITING** on **MATHEMATICS**

2018

For my father, in memoriam

_____

Contents

Introduction

MIRCEA PITICI

This is the ninth volume in our series of remarkable writings on mathematics. The pieces you will read here were initially published during 2017 in various venues, including academic and other professional journals, book chapters, online publications, or newspapers. Except for a few technical mathematical notions required to understand select pieces, the book is accessible to the public that does not specialize in mathematics; yet the book will also interest mathematicians and scientists. Aiming to a wide audience has been, and remains, one of our goals when we prepare every volume.

The origins of *The Best Writing on Mathematics* series go back about fifteen years, to a time when my frustration with the clichés about mathematics I was reading and hearing made me curious to know opinions about mathematics not only from mathematicians but also from outsiders. I quickly discovered that a considerable literature on mathematics authored by mathematicians and by nonmathematicians exists and thrives. Despite its richness in ideas, it is mostly ignored in academic institutions, as if it did not exist and it had no instructional value. For several years, I mulled over the idea of editing such a series, and I attempted to start it; yet life difficulties and the disinterest toward my proposal from the publishers I approached stopped the project in its tracks.

Since 2010, the volumes in this series have contained more than two hundred pieces by authors with diverse backgrounds. These articles range in style from tightly argued theoretical positions on issues related to mathematics to bold speculations on the limits of the applicability of mathematics. Overall, the series is meant to convey to its readers the extraordinary ramifications of the influence of mathematics on contemporary mind, life, and society—and to stimulate connections we usually overlook when we talk about mathematics. The ninth volume is no exception from the general profile of the series.

Overview of the Volume

In the first article in the book, Francis Su gives us an impassioned credo, which he delivered as president of the Mathematical Association of America. Su finds common threads between mathematics on one side and play, beauty, truth, justice, and love on the other side—and from these associations concludes that mathematics contributes to human flourishing.

Margaret Wertheim proposes that certain sentient organisms other than humans, and even nonliving artifacts, perform mathematics whenever they enact rigorous geometric or algebraic patterns—that is, they do so without conscious intelligence.

Robert Thomas considers prevalent views on the "beauty" of mathematics and argues that the quality of making us interested and curious to do mathematics is at least as valuable aesthetically as the quality of pleasing us.

Marijn Heule and Oliver Kullmann write that automated computer proofs are useful and meaningful even if we cannot understand them—and describe how such methods work, detailing one example of proof done by what they call brute reason.

Peter Denning explains how the growth of computational power enabled scientists to change their disciplines from within by adding simulation and information process analysis to the established practices of experimenting and theorizing.

Robbert Dijkgraaf points out that the dynamic of the long interaction between mathematics and physics is reversing; where traditionally mathematics influenced physics, lately physics branches such as string and quantum theories occasion breakthroughs in mathematics, possibly leading toward a new type of mathematics, which he tentatively calls "quantum mathematics."

Erik Demaine and his coauthors describe a planar tangle toy, examine some of the topological configurations available through manipulating it, answer some of the mathematical questions it poses, and formulate a few open problems related to it.

James Grime shows us how to build subtly mischievous dice for playing slightly unfair games!

Arthur Benjamin, Joseph Kisenwether, and Ben Weiss consider another game: bingo. They observe and prove that, contrary to unexamined expectations, bingo winning patterns are asymmetrical, with completed horizontal rows occurring more frequently than completed vertical columns.

Peter Winkler summarizes a plethora of conflicting mathematical, psychological, semantic, and psychological arguments advanced over the past two decades in connection with a question of probability. He purposefully settles nothing, convinced that the controversy surrounding it will continue forever.

José Ferreirós delves into Eugene Wigner's intellectual biography and finds that Wigner's conception of mathematics and some of his consequential epistemological claims were influenced to a considerable degree by his professional friends and associates.

Chris Arney introduces us to the fundamentals of mathematical modeling, including to the necessary theoretical components of modeling, the current and potential scope of applications, and the essential bibliography specialized in mathematical modeling.

Nancy Emerson Kress offers precise advice to school students (and implicitly to teachers, but useful to everyone) on how to approach and to solve problems in mathematics.

On a topic that I have wanted to include in *The Best Writing on Mathematics* anthologies for a long time, Benjamin Braun and his coauthors define active classroom instruction, describe its benefits and variants, and prepare instructors inclined to adopt active teaching methods for some of the challenges and opportunities they are likely to experience.

In a brief note relating more detailed research, Daniel Mansfield and Norman Wildberger explain that in ancient Babylon the mathematicians (or perhaps the surveyors) of the time practiced a different trigonometry than we do today—simpler but precise, founded on the sexagesimal system of numeration.

Isabel Serrano, Lucy Odom, and Bogdan Suceavă examine the structure of mathematics in one of the most important encyclopedic works of premodern times, the *Etymologies* authored by Isidore of Seville in the seventh century.

Michael Barany traces the current awareness of the importance of mathematics in society to developments that started shortly before, but especially during, the Second World War; Barany tells the story of the people who accomplished the major shift in research funding from a long tradition of private sponsorship to one supported in considerable part by the state.

Finally, Caroline Yoon argues that writing and doing mathematics have more in common than we usually admit—and encourages mathematicians (as well as budding mathematicians) to transfer their mathematical skills into writing competencies. Her advice coincides with one of the main goals of *The Best Writing on Mathematics* series.

More Writings on Mathematics

Every year we offer further reading suggestions on mathematics, chosen from recent publications. Toward the end of the book you can find a long list of pieces I considered for selection in this volume. Here, I list some of the books that have come to my attention lately.

I make special mention of the graphically exceptional collection *The Arts of Ornamental Geometry*, edited by Gülru Necipoğlu and the collection of interviews with Russian mathematicians edited by Andrei Sobolevski under the title *Mathematical Walks*.

Among books on mathematics in daily life, including puzzles and games, you can pick from *Gladiators, Pirates and Games of Trust* by Haim Shapira, *Chancing It* by Robert Matthews, *We Are Data* by John Cheney-Lippold, *A Survival Guide to the Misinformation Age* by David Helfand, *The Math Behind . . .* by Colin Beveridge, and *The Power of Networks* by Christopher Brinton and Mung Chiang.

You will find plenty more expository and exciting mathematics at an accessible level in *Foolproof* by Brian Hayes, *Mathematics Rebooted* by Lara Alcock, *Arithmetic* by Paul Lockhart, and even in (the now second volume of) *The Mathematics of Various Entertaining Subjects* edited by Jennifer Beineke and Jason Rosenhouse. More challenging for the mathematically uninitiated, yet appealing to the interested learner with a solid background in school mathematics are *Prime Numbers and the Riemann Hypothesis* by Barry Mazur and William Stein, *Introduction to Experimental Mathematics* by Søren Eilers and Rune Johansen, *Modern Cryptography and Elliptic Curves* by Thomas Shemanske, *A Conversational*

Introduction to Algebraic Theory by Paul Pollack, and the very original and eye-pleasing presentation in *An Illustrated Theory of Numbers* by Martin Weissman.

In the history of mathematics literature, we have two works on China: Daniel Morgan's *Astral Sciences in Early Imperial China*, and Tina Su Lyn Lim and Donald Wagner's *The Continuation of Ancient Mathematics*. For mathematics on other meridians, we cite Jason Fagone's *The Woman Who Smashed Codes* [Elizebeth Smith Friedman], Christopher Graney's *Mathematical Disquisitions* [on Galileo], and Johnny Ball's *Wonders beyond Numbers*. Also in the field of history are *Mathematical Cultures*, edited by Brendan Larvor, *The Mathematics of Secrets* by Joshua Holden, Edward Watts's *Hypatia*, and Michel Serres's *Geometry*. An adventurous autobiography is Ted Hill's *Pushing Limits*.

On statistics, probability, and data, you can learn about *Ten Great Ideas about Chance* from Persi Diaconis and Brian Skyrms, and about *The Tao of Statistics* from Dana Keller. Jeffrey Stanton authored *Reasoning with Data*, and David Carlson published *Quantitative Methods in Archaeology Using R*.

Mathematics is applied in many disciplines, and the literature dedicated to applications is huge. Here are several new titles: *Models, Mathematics, and Methodology in Economic Explanation* by Donald Katzner and *How Much Inequality Is Fair?* by Venkat Venkatasubramanian in economics; *Narrative and Numbers* by Aswath Damodaran, *The Money Formula* by Paul Wilmott and David Orrell, *The Spider Network* by David Enrich, and *Analysing Quantitative Survey Data for Business and Management Students* by Jeremy Dawson in business and finance; *The Birth of Physics* by Michel Serres, *Reality Is Not What It Seems* by Carlo Rovelli, and *It's about Time* by Roger Cooke in physics; *Computational Modeling of Cognition and Behavior* by Simon Farrell and Stephan Lewandowsky, *How Our Days Became Numbered* by Dan Bouk, and *Political Games* by Macartan Humphreys in psychology and other social sciences. A wide-ranging collection of contributions on computing and its history, especially related to Alan Turing's work, is *The Once and Future Turing*, edited by Barry Cooper and Andrew Hodges. In *New Lines*, Matthew Wilson applies mathematical methods to geography and mapmaking, while Paul Charbonneau does so for a multitude of contexts in *Natural Complexity*. An eminently accessible introduction to the mathematics of basic models is *Algebra with Models* by Marian Anton and Karen Santoro. Finally in this category, an idiosyncratic use of mathematics is given by Ben Blatt in *Nabokov's Favorite Word Is Mauve*.

In the philosophy of mathematics, recent titles include *Rethinking Knowledge* by Carlo Cellucci, *Gödel's Disjunction* by Leon Horsten and Philip Welch, *The Ethics of Technology* by Martin Peterson, *Objectivity, Realism, and Proof*, edited by Francesca Boccuni and Andrea Sereni, and *Categories for the Working Philosopher* by Elaine Landry. In the topic of logic are *Bolzano's Conception of Grounding* by Stefan Roski and Kurt Gödel's *Logic Lectures*. A wide-ranging book difficult to categorize is John Stillwell's *Reverse Mathematics*.

Many interdisciplinary books contain mathematical arguments at their hearts. This type of literature is expanding fast. Here are some titles: *Convergence* by Peter Watson, *The Mathematical Corporation* by Josh Sullivan and Angela Zutavern, *Observation and Experiment* by Paul Rosenbaum, *Numbers and the Making of Us* by Caleb Everett, *Mathematics as a Tool* edited by Johannes Lenhard and Martin Carrier, *Autonomous Nature* by Carolyn Merchant, *Scale* by Geoffrey West, *Do the Math!* by John White, *The Probabilistic Foundations of Rational Learning* by Simon Huttegger, and *Game Changers* by Rudolf Taschner.

A book that bears the qualification "dictionary" but in fact is a collection of relevant place descriptions, is *Mathematical Berlin* by Iris and Martin Grötschel. Almost exclusively visual but full of interesting information is *The Book of Circles* by Manuel Lima.

⊚⍦⊚

I hope that you, the reader, will enjoy reading this anthology at least as much as I did working on it. I encourage you to send comments, suggestions, and materials I might consider for (or mention in) future volumes to Mircea Pitici, P.O. Box 4671, Ithaca, NY 14852; or send electronic correspondence to mip7@cornell.edu.

Books Mentioned

Alcock, Lara. *Mathematics Rebooted: A Fresh Approach to Understanding*. Oxford, U.K.: Oxford University Press, 2017.

Anton, Marian, and Karen Santoro. *Algebra with Models: A Guided Inquiry Approach*. San Bernardino, CA: CreateSpace, 2017.

Ball, Johnny. *Wonders beyond Numbers: A Brief History of All Things Mathematical*. London: Bloomsbury, 2017.

Beineke, Jennifer, and Jason Rosenhouse. (Eds.) *The Mathematics of Various Entertaining Subjects*, Vol. 2. Princeton, NJ: Princeton University Press, 2017.

Beveridge, Colin. *The Math Behind. . .: Discover the Mathematics of Everyday Events*. Richmond Hill, Canada: Firefly Books.

Blatt, Ben. *Nabokov's Favorite Word Is Mauve: What the Numbers Reveal about the Classics, Bestsellers, and Our Own Writing*. New York: Simon and Schuster, 2018.

Boccuni, Francesca, and Andrea Sereni. (Eds.) *Objectivity, Realism, and Proof.* Cham, Switzerland: Springer International Publishing, 2017.

Bouk, Dan. *How Our Days Became Numbered: Risk and the Rise of the Statistical Individual*. Chicago: University of Chicago Press, 2018.

Brinton, Christopher G., and Mung Chiang. *The Power of Networks: Six Principles That Connect Our Lives*. Princeton, NJ: Princeton University Press, 2017.

Carlson, David L. *Quantitative Methods in Archaeology Using R*. Cambridge, U.K.: Cambridge University Press, 2017.

Cellucci, Carlo. *Rethinking Knowledge: The Heuristic View*. Cham, Switzerland: Springer Science+Business Media, 2017.

Charbonneau, Paul. *Natural Complexity: A Modeling Handbook*. Princeton, NJ: Princeton University Press, 2017.

Cheney-Lippold, John. *We Are Data: Algorithms and the Making of Our Digital Selves*. New York: New York University Press, 2017.

Cooke, Roger. *It's about Time: Elementary Mathematical Aspects of Relativity*. Providence, RI: Mathematical Association of America, 2017.

Cooper, S. Barry, and Andrew Hodges. (Eds.) *The Once and Future Turing: Computing the World*. Cambridge, U.K.: Cambridge University Press, 2016.

Damodaran, Aswath. *Narrative and Numbers: The Value of Stories in Business*. New York: Columbia Business School Publishing, 2017.

Dawson, Jeremy. *Analysing Quantitative Survey Data for Business and Management Students*. Thousand Oaks, CA: Sage, 2017.

Diaconis, Persi, and Brian Skyrms. *Ten Great Ideas about Chance*. Princeton, NJ: Princeton University Press, 2017.

Eilers, Søren, and Rune Johansen. *Introduction to Experimental Mathematics*. Cambridge, U.K.: Cambridge University Press, 2017.

Enrich, David. *The Spider Network: The Wild Story of a Math Genius, a Gang of Backstabbing Bankers, and One of the Greatest Scams in Financial History*. New York: HarperCollins, 2017.

Everett, Caleb. *Numbers and the Making of Us: Counting and the Course of Human Cultures*. Cambridge, MA: Harvard University Press, 2017.

Fagone, Jason. *The Woman Who Smashed Codes: A True Story of Love, Spies, and the Unlikely Heroine Who Outwitted America's Enemies* [Elizebeth Smith Friedman]. New York: HarperCollins, 2017.

Farrell, Simon, and Stephan Lewandowsky. *Computational Modeling of Cognition and Behavior*. Cambridge, U.K.: Cambridge University Press, 2018.

Gödel, Kurt. *Logic Lectures: Gödel's Basic Logic Course at Notre Dame*. Belgrade, Serbia: Logical Society of Belgrade, 2017.

Graney, Christopher M. *Mathematical disquisitions: The Booklet of Theses Immortalized by Galileo*. Notre Dame, IN: University of Notre Dame Press, 2017.

Grötschel, Iris, and Martin Grötschel. *Mathematical Berlin: Science, Sights, and Stories*. Berlin, Germany: Berlin Story Verlag, 2016.

Hayes, Brian. *Foolproof, and Other Mathematical Meditations*. Cambridge, MA: MIT Press, 2017.

Helfand, David J. *A Survival Guide to the Misinformation Age: Scientific Habits of Mind.* New York: Columbia University Press, 2017.

Hill, Ted. *Pushing Limits: From West Point to Berkeley and Beyond.* Providence, RI: American Mathematical Society, and Washington, DC: Mathematical Association of America, 2017.

Holden, Joshua. *The Mathematics of Secrets: Cryptography from Caesar Cyphers to Digital Encryption.* Princeton, NJ: Princeton University Press, 2017.

Horsten, Leon, and Philip Welch. *Gödel's Disjunction: The Scope and Limits of Mathematical Knowledge.* New York: Oxford University Press, 2016.

Humphreys, Macartan. *Political Games: Mathematical Insights on Fighting, Voting, Lying and Other Affairs of State.* New York: W. W. Norton, 2016.

Huttegger, Simon M. *The Probabilistic Foundations of Rational Learning.* Cambridge, U.K.: Cambridge University Press, 2017.

Katzner, Donald W. *Models, Mathematics, and Methodology in Economic Explanation.* Cambridge, U.K.: Cambridge University Press, 2018.

Keller, Dana K. *The Tao of Statistics: A Path to Understanding (with No Math).* Los Angeles: Sage, 2017.

Landry, Elaine. *Categories for the Working Philosopher.* Oxford, U.K.: Oxford University Press, 2017.

Larvor, Brendan. (Ed.) *Mathematical Cultures: The London Meetings 2012–2014.* Cham, Switzerland: Springer International Publishing, 2016.

Lenhard, Johannes, and Martin Carrier. (Eds.) *Mathematics as a Tool: Tracing New Roles of Mathematics in the Sciences.* Cham, Switzerland: Springer International Publishing, 2017.

Lim, Tina Su Lyn, and Donald B. Wagner. *The Continuation of Ancient Mathematics: Wang Xiaotong's Jigu suanjing, Algebra and Geometry in 7th-Century China.* Copenhagen, Denmark: Nordic Institute for Asian Studies, 2017.

Lima, Manuel. *The Book of Circles: Visualizing Spheres of Knowledge.* Princeton, NJ: Princeton Architectural Press, 2017.

Lockhart, Paul. *Arithmetic.* Cambridge, MA: Harvard University Press, 2017.

Matthews, Robert. *Chancing It: The Laws of Chance and How They Can Work for You.* London: Profile Books, 2017.

Mazur, Barry, and William Stein. *Prime Numbers and the Riemann Hypothesis.* Cambridge, U.K.: Cambridge University Press, 2016.

Merchant, Carolyn. *Autonomous Nature: Problems of Prediction and Control from Ancient Times to the Scientific Revolution.* Abingdon, U.K.: Routledge, 2016.

Morgan, Daniel Patrick. *Astral Sciences in Early Imperial China: Observation, Sagehood and the Individual.* Cambridge, U.K.: Cambridge University Press, 2017.

Necipoğlu, Gülru. (Ed.) *The Arts of Ornamental Geometry: A Persian Compendium on Similar and Complementary Interlocking Figures.* Leiden, Netherlands: Brill, 2017.

Peterson, Martin. *The Ethics of Technology: A Geometric Analysis of Five Moral Principles.* New York: Oxford University Press, 2017.

Pollack, Paul. *A Conversational Introduction to Algebraic Theory: Arithmetic beyond Z.* Providence, RI: American Mathematical Society, 2017.

Rosenbaum, Paul R. *Observation and Experiment: An Introduction to Causal Inference.* Cambridge, MA: Harvard University Press, 2017.

Roski, Stefan. *Bolzano's Conception of Grounding.* Frankfurt am Main, Germany: Vittorio Klostermann, 2017.

Rovelli, Carlo. *Reality Is Not What It Seems: The Journey to Quantum Gravity.* New York: Random House, 2017.

Serres, Michel. *The Birth of Physics*. London, UK: Rowman & Littlefield, 2018 [French orig. 1977].

Serres, Michel. *Geometry: The Third Book of Foundations*. London: Bloomsbury Academic, 2017.

Shapira, Haim. *Gladiators, Pirates and Games of Trust: How Game Theory, Strategy and Probability Rule Our Lives*. London: Watkins, 2017.

Shemanske, Thomas R. *Modern Cryptography and Elliptic Curves: A Beginner's Guide*. Providence, RI: American Mathematical Society, 2017.

Sobolevski, Andrei. (Ed.) *Mathematical Walks: A Collection of Interviews*. Moscow: Skolkovo Institute of Science and Technology, 2017.

Stanton, Jeffrey M. *Reasoning with Data: An Introduction to Traditional and Bayesian Statistics Using R*. New York: Guilford Press, 2017.

Stillwell, John. *Reverse Mathematics: Proofs from the Inside Out*. Princeton, NJ: Princeton University Press, 2017.

Sullivan, Josh, and Angela Zutavern. *The Mathematical Corporation: Where Machine Intelligence and Human Ingenuity Achieve the Impossible*. New York: Perseus Books, 2017.

Taschner, Rudolf. *Game Changers: Stories of the Revolutionary Minds behind Game Theory*. Amherst, NY: Prometheus Books, 2017.

Venkatasubramanian, Venkat. *How Much Inequality Is Fair? Mathematical Principles of a Moral, Optimal, and Stable Capitalist Society*. New York: Columbia University Press, 2017.

Watson, Peter. *Convergence: The Idea at the Heart of Science*. New York: Simon and Schuster, 2016.

Watts, Edward J. *Hypatia: The Life and Legend of an Ancient Philosopher*. Oxford, U.K.: Oxford University Press, 2017.

Weissman, Martin H. *An Illustrated Theory of Numbers*. Providence, RI: American Mathematical Society, 2017.

West, Geoffrey. *Scale: The Universal Laws of Growth, Innovation, Sustainability, and the Pace of Life in Organisms, Cities, Economies, and Companies*. New York: Penguin Press, 2017.

White, John K. *Do the Math! On Growth, Greed, and Strategic Thinking*. Los Angeles, CA: Sage, 2013.

Wilmott, Paul, and David Orrell. *The Money Formula: Dodgy Finance, Pseudo Science, and How Mathematicians Took Over the Markets*. Chichester, U.K.: John Wiley & Sons, 2017.

Wilson, Matthew W. *New Lines: Critical GIS and the Trouble of the Map*. Minneapolis: University of Minnesota Press, 2017.

The BEST WRITING on MATHEMATICS

2018

Mathematics for Human Flourishing

Francis Edward Su

Every being cries out silently to be read differently.
—Simone Weil [7, p. 188]

Christopher is an inmate in a high-security federal prison not far from Atlanta. He's been in trouble with the law since he was fourteen. He didn't finish high school, had an addiction to hard drugs, and at age twenty-one, his involvement in a string of armed robberies landed him in prison with a thirty-two-year sentence.

Right now, you've probably formed a mental image of who Christopher is, and you might be wondering why I'm opening my speech with his story. When you think about who does mathematics—both who is capable of doing mathematics and who wants to do mathematics—would you think of Christopher?

And yet he wrote me a letter after seven years in prison. He said, "I've always had a proclivity for mathematics, but being in a very early stage of youth and also living in some adverse circumstances, I never came to understand the true meaning and benefit of pursuing an education . . . over the last 3 years I have purchased and studied a multitude of books to give me a profound and concrete understanding of Algebra I, Algebra II, College Algebra, Geometry, Trigonometry, Calculus I and Calculus II."

Christopher was writing me for help in furthering his mathematics education.

When you think of who does mathematics, would you think of Christopher?

Every being cries out silently to be read differently.

Simone Weil is a well-known French religious mystic and a widely revered philosopher. She is probably less well known as the younger sister of André Weil, one of history's most famous number theorists.

For Simone, to *read* someone means to interpret or make a judgment about them. She's saying, "Every being cries out silently to be judged differently." I sometimes wonder if Simone was crying out about herself. For she, too, loved and participated in mathematics, but she was always comparing herself to her brother. She wrote [6, p. 64],

> At fourteen I fell into one of those fits of bottomless despair that come with adolescence, and I seriously thought of dying because of the mediocrity of my natural faculties . . . the exceptional gifts of my brother, who had a childhood and youth comparable to those of Pascal, brought my own inferiority home to me. I did not mind having no visible successes, but what did grieve me was the idea of being excluded from that transcendent kingdom to which only the truly great have access and wherein truth abides. I preferred to die rather than live without that truth.

We know Simone loved mathematics because she used mathematical examples in her philosophical writing. And you'll find her in photos of Bourbaki with her brother.

I often wonder what her relationship to mathematics would be like if she weren't always in André's shadow.

Every being cries out silently to be read differently.

As president of the Mathematical Association of America (MAA), you might think that my relationship to mathematics has always been solid. I don't like the word "success," but people look at me and think I'm successful, as if the true measure of mathematical achievement is the grants I've received or the numerous papers I've published.

Like Christopher, I've had a proclivity for mathematics since youth. But I grew up in a small rural town in South Texas, with limited opportunities. Most of my high school peers didn't even attend college. I did because my dad was a college professor, but my parents didn't know about the many mathematical opportunities I now know exist.

My love for math deepened at the University of Texas, and I managed to get admitted to Harvard for my Ph.D. But I felt out of place there, since I did not come from an Ivy League school, and unlike my peers, I did not have a full slate of graduate courses when I entered. I felt like Simone Weil, standing next to future André Weils, thinking I would never be able to flourish in mathematics if I were not like them.

I was told by one professor that I didn't belong in graduate school. That forced me to consider, among other things, why I wanted to do mathematics. And in fact, that is essentially the one big question that I'd like for you to consider today.

Why do mathematics?

This is a simple question, but worth considerable reflection. Because how you answer will strongly determine *who* you think should be doing mathematics, and *how* you will teach it.

Why is Christopher sitting in a prison cell studying calculus, even though he won't be using it as a free man for another twenty-five years? Why was Simone so captivated by transcendent mathematical truths? Why should anyone persist in doing math or seeing herself as a mathematical person when others are telling her in subtle and not so subtle ways that she doesn't belong?

And in this present moment, the world is also asking what its relationship with mathematics should be. Amid the great societal shifts wrought by the digital revolution and a shift to an information economy, we are witnessing the rapid transformation of the ways we work and live. And yet we hear voices in the public sphere, saying "high school students don't need geometry" or "let's leave advanced math for the mathematicians." And some mathematicians won't admit it, but they signal exactly the same thing by refusing to teach lower-level math courses or viewing the math major as a means to weed out those they don't think are fit for graduate school.

Our profession is threatened by voices like these from within, and without, who are undermining how society views mathematics and mathematicians. And the view of our profession is dismal. The 2012 report from the President's Council of Advisors on Science and Technology pegs introductory math courses as the major obstacle keeping students from pursuing STEM majors. We are not educating our students as well as we should, and like most injustices, this hurts those who are most vulnerable.

I want us as a mathematical community to move forward in a different way. It may require us to change our view of who should be doing mathematics and how we should teach it. But this way will be no less rigorous and no less demanding of our students. And yet it will draw more people into mathematics because they will see how mathematics connects to their deepest human desires.

So if you asked me, "Why do mathematics?" I would say, "Mathematics helps people flourish."

Mathematics is for human flourishing.

The well-lived life is a life of human flourishing. The ancient Greeks had a word for human flourishing, *eudaimonia*, which they viewed as the good composed of all goods. There is a similar word in Hebrew: *shalom*, which is used as a greeting. Shalom is sometimes translated "peace," but the word has a far richer context. To say "shalom" to someone is to wish that they will flourish and live well. And Arabic has a related word: *salaam*.

A basic question, taken up by Aristotle, is this: How do you achieve human flourishing? What is the well-lived life? Aristotle would say that flourishing comes through the exercise of virtue. The Greek concept of virtue is excellence of character that leads to excellence of conduct. So it includes more than just moral virtue; for instance, courage and wisdom are also virtues.

What I hope to convince you of today is that the practice of mathematics cultivates virtues that help people flourish. These virtues serve you well no matter what profession you choose. And the movement toward virtue happens through basic human desires.

I want to talk about five desires we all have. The first of these is play.

1. PLAY

It is a happy talent to know how to play.
—Ralph Waldo Emerson [2, p. 138]

Think of how babies play. Play is hard to define, but we can think of a few qualities that characterize it. For instance, play should be *fun* and *voluntary*, or it wouldn't be play. There is usually some *structure*—even babies know that "peekaboo" follows a certain pattern—but there is lots of *freedom* within that structure. That freedom leads to investigation of some sort, like "where will you appear if we play peekaboo one more time?" There is usually *no great stake* in the outcome. And the investigation can often lead to some sort of *surprise*, like appearing in a different place in peekaboo. Of course, animals play too, but what characterizes human play is the enlarged role of mind and the *imagination*.

Think about Rubik's Cube or the game Set. There's interplay between structure and freedom and no great stake in the outcome, but there's investigation that can lead to the delight of solving the cube or finding sets of matching cards.

Mathematics makes the mind its playground. We play with patterns, and within the structure of certain axioms, we exercise freedom in exploring their consequences, joyful at any truths we find. We even have a whole area of mathematics known as "recreational mathematics"! Do you know another discipline that has a "recreational" subfield? Is there a "recreational physics" or "recreational philosophy"?

And mathematical play builds virtues that enable us to flourish in every area of our lives. For instance, math play builds **hopefulness**—when you sit with a puzzle long enough, you are exercising hope that you will eventually solve it. Math play builds **community**—when you share in the delight of working on a problem with another human being. And math play builds **perseverance**—just as weekly soccer practices build up the muscles that make us stronger for the next game, weekly math investigations make us more fit for the next problem, whatever that is, even if we don't solve the current problem. It's why the MAA supports programs like the American Mathematics Competitions and the Putnam Competition. We help kids flourish through building hopefulness, perseverance, and community. This year, you may have heard that the U.S. team, which MAA trained, won the International Math Olympiad for the second time in a row. What you might not have heard is that Po-Shen Loh, who coached our team, invited teams from other countries to train with them to prepare. You see, our priority was community over competition. This action was so impressive to the Singaporean prime minister that he publicly thanked President Obama for this remarkable collaboration. This was true play: teams in friendly competition.

Play is part of human flourishing. You cannot flourish without play.

And if mathematics is for human flourishing, we should "play up" the role of play in how we teach and who we teach. Everyone can play. Everyone enjoys play. Everyone can have a meaningful experience in mathematical play.

And teaching play is hard work! It's actually harder than lecturing because you have to be ready for almost anything to happen in the classroom, but it's also more fun. Play is part of what makes inquiry-based

learning and other forms of active learning so effective. There's over-
whelming evidence that students learn better with active learning. This
year, in the Conference Board of the Mathematical Sciences, I signed a
statement with presidents of other math organizations endorsing active
learning, available on the CBMS website. And if you want to see the
evidence for active learning, we've included some background information in this statement.

So, teach play.

Another basic human desire is beauty.

2. BEAUTY

It is impossible to be a mathematician
without being a poet in soul.
—Sónya Kovalévsky [3, p. 316]

Who among us does not enjoy beautiful things? A beautiful sunset. A
sublime sonata. A profound poem. An elegant proof.

Mathematicians and scientists are awed by the simplicity, regularity, and order of the laws of the universe. These are called "beautiful."
They feel transcendent. Why should mathematics be as powerful as it
is? This is what Nobel prize–winning physicist Eugene Wigner called
"the unreasonable effectiveness of mathematics" to explain the natural
sciences. And Einstein asked, "How can it be that mathematics, being
after all a product of human thought independent of experience, is so
admirably adapted to the objects of reality?"

And mathematicians are not satisfied with just any proof of a theorem. We often look for the best proofs, the simplest or most pleasing.
Mathematicians have a special word for this—we say a proof is "elegant." Paul Erdős often spoke of "The Book" that God keeps, in which
all the most elegant proofs of theorems are recorded.

Pursuing mathematics in this way cultivates the virtues of **transcendence** and **joy**. By joy, I refer to the wonder or awe or delight in
the beauty of the created order. By transcendence, I mean the ability to
embrace the mystery of it all. There's a transcendent joy in experiencing the beauty of mathematics.

If mathematics is for human flourishing, we must help others see its beauty.

But there are many notions of beauty. So the way you motivate mathematics through beauty must necessarily be diverse—through art, through music, through patterns, through rigorous arguments, through the elegance of simple but profound ideas, through the wondrous applicability of these ideas to the real world in many different fields.

A third basic human desire is truth.

3. TRUTH

Quid est veritas?
—*Pontius Pilate [5, p. 153]*

What is truth? This question is an important one, especially today. Each day seems to bring more discussion about how fake news may have influenced the 2016 presidential election. Some dismiss trying to figure out what's true, saying, "How can we even know what is true?"

And yet, in some contexts, people do seek truth at all costs, especially when there is a lot at stake. When my dad had cancer, we wanted to know what treatments had the best chance of saving his life. We had to know because his life depended on it.

The quest for truth is at the heart of the scientific enterprise. I say "quest" because we don't do science to confirm simple declarative statements that are easily verified, like "my coffee is hot." Rather, our subjects of investigation are questions for which the answer is not so clear. "Do gravity waves exist, and if so, how would we detect them?"

So there is a quest. We formulate a hypothesis ("they exist"), and we design experiments to test our hypothesis. We look for evidence, and if we find some, we still ask, "Could there have been any other explanation?"

A mathematician might try to prove or disprove a statement through logical deductions from first principles. Or she might construct a mathematical model to answer the question.

These approaches cultivate in us the virtue of **rigorous thinking**: the ability to handle ideas well and to craft clear arguments with those ideas. This virtue serves us well in every area of life. We should use this ability to reason in the public square, as many in our community have done by writing op-ed pieces for newspapers. More of us should exercise this virtue to shape public perceptions about mathematics.

I would like to encourage institutions to start valuing the public writing of its faculty. More people will read these pieces than will ever read any of our research papers. Public writing is scholarly activity. It involves rigorous arguments, is subject to a review process by editors, and to borrow the phrase from the National Science Foundation, it has broader impacts, and that impact can be measured in the digital age.

The quest for truth predisposes the heart to the virtue of **humility**. Isaac Newton said, "I do not know what I may appear to the world, but to myself I seem to have been only like a boy playing on the seashore, and diverting myself in now and then finding a smoother pebble or a prettier shell than ordinary, whilst the great ocean of truth lay all undiscovered before me" [1, p. 407]. He's saying that the more we know, the more we realize how much we do not yet know. And we learn how to accept being wrong if a counterexample shows that our conjecture was false. In fact, I'll go so far as to say that counterexamples in mathematics have a special place—we celebrate them. We have titles of books like "Counterexamples in Topology" or "Counterexamples in Analysis." We like to admit when we are wrong!

So when a student embraces this quest for truth, she begins to assume a certain kind of humility. She handles ideas rigorously, with honesty and integrity. She values truthfulness and clarification of distinctions. This is the virtue of intellectual humility, and it is prized. I think a lack of humility characterized the political discourse of 2016 on both sides. I wish we had more intellectual humility in the public square.

We must model the virtue of humility in our own teaching, and we should explicitly tell our students we are cultivating humility as a virtue that will serve them well their entire lives. One of the most important skills we can teach our students is to know when their arguments are wrong. How many of you have ever given a super-hard question on an exam and gotten answers that look like students just made stuff up, hoping for partial credit? I now explicitly say on my exams that I will give extra credit on incomplete proofs where students acknowledge their gaps. I get much more thoughtful answers that way.

And mathematics builds the virtue of **circumspection**. We know the limits of our arguments, and we don't overgeneralize. I like what my friend Rachel Schwell said:

"I think math helps me make fewer sweeping generalizations about people. For example, I wouldn't assume a person is, say, uneducated, just because she is, say, poor, just as I can't assume a number is say, positive, because it is an integer. I can't even assume it is positive if I know it's nonnegative, even if, probability-wise, it probably is positive! So I don't leap to automatic associations as much."

Can we help our students see that the virtue of circumspection is important in life?

A fourth basic human desire is justice.

4. Justice

Justice. To be ever ready to admit that another person is
something quite different from what we read when he is there
(or when we think about him). Or, rather, to read in him
that he is certainly something different, perhaps something
completely different from what we read in him.
Every being cries out silently to be read differently.
—Simone Weil [7, p. 188]

Akemi was a student of mine who did research with me as an undergraduate. Her innovative paper linking game theory and phylogenetics was published in a highly regarded mathematical biology journal. She went to a top research university for her Ph.D. So I was surprised when I learned that Akemi quit after one year.

She told me that she had many negative experiences. Her advisor was never willing to meet with her, and she had faced uncomfortable experiences as a woman. She told me one example:

At the beginning of the course, I consistently got 10/10 on my homework assignments which were all graded by the TA. One day, Jeff [a mutual friend] told me that he was hanging out with our TA and someone asked the TA how the analysis class was doing. He went on and on about some "guy" named Akemi and how perfect "his" homeworks were and how clearly they were written, etc. Jeff told him I was a girl and the TA was shocked. (Jeff told me this story because he thought it was funny that someone both didn't know my sex from my name and reacted so dramatically to finding out.) After that, I never got remotely close to 10/10 on my assignments and my exams were equally harsh—most of the reasons for docked points were vague with comments like "give more detail." I didn't feel like my understanding of the material diminished that quickly or dramatically.

I hope you agree something is not right with this picture. If a certain kind of anger wells up in you, you are experiencing a telltale sign of flourishing: the desire for justice. Justice means setting things right. And justice is a powerful motivator to action.

Justice is required for human flourishing. We flourish—we experience shalom—when we treat others justly and when we are treated justly.

Simone Weil realized that correcting injustice must involve changing how we view others: "to read in him that he is certainly something different, perhaps something completely different from what we read in him. Every being cries out silently to be read differently."

Now before we are too quick to censure Akemi's TA, we have to realize that the problem of reading others differently begins with ourselves. The TA may not have even realized he was doing this. This is the problem of *implicit bias*: unconscious stereotypes that subtly affect our decisions. One of the best experiences I had in MAA leadership was attending a workshop on implicit bias, in which I realized in a powerful way how I am biased even though I try not to be. We all do it without realizing it. Numerous experiments confirm results of the following kind: When given two nearly identical resumes except that one has a positively stereotyped name and one has a negatively stereotyped name (e.g., woman or minority), judges rate the positively stereotyped resume higher. This happens even if judges come from the negatively stereotyped group.

This is why good practices are important. The MAA now has a document for selection committees called "Avoiding Implicit Bias" that lists a number of practices that have been shown in research to mitigate the effects of implicit bias, such as taking time to make decisions or generating a large candidate pool. These are good practices even if you don't believe that bias exists. That document is now distributed with every MAA committee assignment.

You see, we have to recognize that even if people are just, even if they desire to be just, a society may not be just if its structures and practices are not also just. And the only way a whole society can flourish is if the society is a just society. It is often said that the mark of a just society is how it treats its most vulnerable members.

So I ask, with great humility, are we a just community?

If you believe that mathematics is for human flourishing, and we teach mathematics to help people flourish, you will see, if you look around the room, that we aren't helping all our students flourish. The demographics of the mathematical community do not look like the demographics of America. We have left whole segments out of the benefits of the flourishing available in our profession.

So we have to talk about race, and that's hard. It can bring up complicated emotions, even more so with all that has taken place in our nation in the past year. In our community, we have to become more comfortable talking about race, listening to each other's experiences, and being willing to recognize it's there. If you want to treat others with dignity and they are hurting, you don't ignore their pain. You ask, "What are you going through?"

It's not enough to say, "I don't think about race" because in a community, how one member is doing affects the whole community. And for those of us not in the dominant racial group, we don't have the luxury of saying, "I don't think about race" because racial issues affect us on a daily basis. So let me encourage all of us to try having these conversations, to be quick to listen, slow to speak, and quick to forgive each other when we say something stupid. That'll happen if you start to have conversations, and we just have to have grace for each other if we make mistakes—it's better than not talking.

So if we're going to have conversation, I'll start. I grew up in Texas in a white and Latino part of the state, and I realized early on that my family had different customs from my friends—my clothes were different, the food in my lunchbox was different—and these things were causing me not to fit in. I wanted to be white. Not Latino, because white people got more respect, and as an Asian, I was getting picked on all the time. I had no role models for being Asian-American. So I tried hard to act white, even if I couldn't look white.

On the other hand, in Chinese communities, I also don't fit in. I don't speak Chinese. I don't act Chinese. At Chinese restaurants, I'm viewed as white. Did you know that at authentic Chinese restaurants, there is often a special menu, a secret menu, that they only give to Chinese people? It has all the good stuff. I don't get that menu unless I ask for it. In fact, they discourage me, saying, "You won't like the stuff on that menu."

As mathematicians, who gets to see our secret menu? Whom do we shepherd toward taking more math courses? Whom do we discourage from looking at that menu?

Don't let me sound as if I'm complaining about my race. There are ways in which I benefited from being Asian. People expected me to do better at math and science, and I'm sure that's part of why I did. Because I now know there is a recognized literature on "expectancy effects," that teacher expectations do affect student performance.

The first time I didn't feel like a minority was when I moved to California. There are so many Asian-Americans there. In Texas, I would commonly get the question, "Your English is so good! Where are you from?"

"Texas."

"No, where are you *really* from?" That never happens in California, and there's a feeling of freedom I have in not having to counter these verbal stings.

These days, I'm used to being at math conferences and seeing a sea of white faces. So even I was a little bit surprised that when I was elected MAA president, a prominent blogger on race issues for Asian-Americans wrote a blog post about it. His name is Angry Asian Man. He looked at the photos of past MAA presidents on our website, and given how many Asians he expected to be in math, he noted that they were all white except for me and wrote a sarcastic post entitled

"Finally, an Asian guy who's good at math."

I am the first president of color of either the American Mathematical Society (AMS) or MAA. Minorities, including Asians, are easy to overlook when you think about who would make a good leader. This situation may not be intentional, but when you are asked to think about who is fit for this or that role, you often think of people just like the people who have been in office. So it is easy for implicit bias to creep in.

I raise this discussion out of deep affection for the mathematical community. I want us to flourish, and there are ways in which we can do better.

In 2015, I had the great pleasure of running MSRI-UP, a summer research program for students from underrepresented backgrounds: first-generation college students, Latino and African-American kids. I asked them to help me prepare this talk, to tell me about obstacles they've faced doing mathematics.

One of them, who did wonderful work that summer, told me about her experience in an analysis course after she got back. She said, "Even though the class was really hard, it was more difficult to receive the humiliations of the professor. He made us feel that we were not good enough to study math and he even told us to change to another 'easier' profession." As a result of this and other experiences, she switched her major to engineering.

Let me be clear: There is no good reason to tell a student she doesn't belong in math. That's the student's decision, not yours. You see the snapshot of her progress, but you don't see her trajectory. You can't know how she will grow and flourish in the future. But you can help her get there.

Of course, you should give forthright counsel to students about skills they might need to develop further if they want to go on in mathematics, but if you see mathematics as a means to help them flourish, why wouldn't you encourage them to take more math?

Another student that summer, Oscar, told me about his experience as a math major. Unlike his peers and because of his background, he did not enter college with any advanced placement credit. He says,

> I noticed how different my trajectory was, however, while I was in my Complex Analysis course. A student was presenting a solution on the board which required a bit of a complicated derivation halfway through. They skipped over a number of steps, citing "I don't think I need to go through the algebra. . . . we all tested out of Calculus here anyway!" with my professor nodding in agreement and some students laughing. I quietly commented that Calculus

was my first course here. My professor was genuinely surprised and said, "Wow, I did not know that! That's interesting." I was not sure whether to feel proud or embarrassed by the fact that I was not the "typical math student" that was successful from the beginning of their mathematical career. I felt a sense of pride in knowing that I was pursuing a math degree despite my starting point, but I could not help but feel as though I did not belong in that classroom to begin with.

The reason Oscar was in that class to begin with was because of the active support of another professor. Says Oscar,

She presented me with my first research opportunity and always encouraged me to study higher math. I was also able to confide in her about a lot of the internal struggles I had with being a minority in mathematics since, as a female, she had a similar experience herself! My Complex Analysis professor became one of my mentors as well. I think it was just an interesting moment because she didn't realize how her reaction to the situation could have hurt me (and I don't think she's necessarily at fault!). It was more that her reaction piled onto the insecurities I held in regards to being a minority with a weak background in math.

Note that Oscar didn't have a "weak" background—he had a standard background.

I'm pleased to say that Oscar and his team from that summer just published a paper in an AMS journal, and he is now in graduate school.

You hear from Oscar's story the importance of having an advocate, a faculty member who says, "I see you, and I think you have a future in math." This support can be especially important for underrepresented groups and women, who already have so many voices telling them they can't. Can you be that advocate?

And if we teach mathematics to help our students flourish, then we should not set up structures that disadvantage smart students with weaker backgrounds or make them feel out of place. I know that can happen inadvertently among students, but we, as faculty, are the shepherds of our departmental culture. When I was a grad student at Harvard, they had a regular calculus class, an honors calculus class called Math 25, and on top of that, for those with very strong backgrounds, a super honors class called Math 55. Ironically, I regularly encountered students in the *honors* track who felt that they didn't belong in math because they hadn't placed into the super honors track. I had to keep reassuring them that "background is not the same as ability." I sometimes wish graduate school admissions would remember this too: "Background is not the same as ability." As my friend Bill Velez says, "If you want your Ph.D. program to have more students of color, then you have to stop admitting students on the basis of background and start admitting students by their ability. And then, support them." They have so many voices telling them that they don't belong. Be an advocate!

I know our community wants to be just, to set things right. So if you are looking to start some conversations with your students or colleagues but don't know where to start, sometimes it can help to have a third party. I'm willing to be that third party. I've written a number of articles on these topics for *MAA FOCUS*, and they are all posted on my web page. You could ask your students to read them and then have a discussion. I can assure you that it will be time well spent.

For we are not mathematical machines; we live, we breathe, we feel, we bleed. If your students are struggling and you don't acknowledge it, their education becomes disconnected and irrelevant. Why should anyone care about mathematics if it doesn't connect deeply to some human desire: to play, seek truth, pursue beauty, or fight for justice? You can be that connection.

So let me challenge each one of you today. Find one student whom you know is facing some challenges, and become that student's long-term advocate. One way to do that is to sign up to be a mentor with the Math Alliance. The goal of this program, directed by Phil Kutzko, is to ensure that every underrepresented or underserved U.S. student with the talent and the ambition has the opportunity to earn a doctoral degree in a mathematical science.

Find one student and be her advocate! Be the one who says, "I see you, and I think you have a future in math." Be the one who searches out opportunities for her. Be the one who pulls her toward virtue. Be the one who calls her when she has skipped class and asks, "Is everything okay? What are you going through?"

I know what I'm asking you to do is hard and takes time.

But we're mathematicians; we know how to tackle hard problems. We have the perseverance to see it through. We have the humility to admit when we make mistakes and to learn from them. We have hopefulness that our labor is never in vain and the transcendent belief that our work will bear fruit in the flourishing of our students.

Because what I am asking you to do is something you already know, at the heart of the teacher–student relationship, pulls us toward virtue.

I'm asking you to love.

5. LOVE

If I speak in the tongues of mortals and of angels, but have not love, I am a noisy gong or a clanging cymbal.
—Paul the Apostle [4, p. 2017]

Love is the greatest human desire. And to love and be loved is a supreme mark of human flourishing. For it serves the other desires—play, truth, beauty, and justice—and it is served by them.

Every being cries out silently to be read differently. Every being cries out silently to be loved. Christopher, in prison, wasn't looking only for mathematical advice. He was looking for connection, someone to reach out to him in his mathematical space and say, "I see you, and I share the same transcendent passion for math that you do, and you belong here."

When I was in the depths of despair in graduate school, struggling over many nonacademic things with a professor who had said I don't belong, already interviewing for jobs because I was sure I was going to quit, one professor reached out to me, became my advocate. And he said, "I would rather see you work with me than quit." So now I stand before you to ask you to find a struggling student, love that student, be his advocate!

I'll close with this reflection by Simone Weil. After wrestling with her own insecurity in mathematics, she saw that there was a path to virtue through her struggle and that her struggle could help others. She wrote [6, pp. 115–116],

The love of our neighbour in all its fullness simply means being able to say to him: "What are you going through?" It is a recognition that the sufferer exists, not only as a unit in a collection, or a specimen from the social category labelled "unfortunate", but as a man, exactly like us, who was one day stamped with a special mark by affliction. For this reason it is enough, but it is indispensable, to know how to look at him in a certain way.

This way of looking is first of all attentive. The soul empties itself of all its own contents in order to receive into itself the being it is looking at, just as he is, in all his truth.

Only he who is capable of attention can do this.

So it comes about that, paradoxical as it may seem, a Latin prose or a geometry problem, even though they are done wrong, may be of great service one day, provided we devote the right kind of effort to them. Should the occasion arise, they can one day make us better able to give someone in affliction exactly the help required to save him, at the supreme moment of his need.

For an adolescent, capable of grasping this truth and generous enough to desire this fruit above all others, studies could have their fullest spiritual effect, quite apart from any particular religious belief.

Academic work is one of those fields which contain a pearl so precious that it is worthwhile to sell all our possessions, keeping nothing for ourselves, in order to be able to acquire it.

Simone Weil had found a path through struggle to virtue. She understood that mathematics is for human flourishing. And the mathematical experience cannot be separated from love:

The love between friends who play with a mathematical problem.

The love between teacher and student growing together toward virtue.

The love of a community like the Mathematical Association of America working with each other toward a common goal: through the knowledge and virtues wrought by mathematics, to help everyone flourish.

Thank you for the opportunity to serve you these last two years. Shalom and salaam, my friends. Grace and peace to you. May you and all your students flourish.

Acknowledgments

I'm grateful for many friends and students who shared their stories and insights with me. And I'm indebted to David Vosburg and Matt

DeLong, whose sustained encouragement, friendship, and wise counsel helped shape this message.

References

1. D. Brewster, *Memoirs of the Life, Writings, and Discoveries of Sir Isaac Newton*, Vol. II. Thomas Constable, Edinburgh, 1855.

2. R. W. Emerson, *Emerson in His Journals*. Ed. J. Porte. Harvard Univ. Press, Cambridge, MA, 1982.

3. S. Kovalévsky, *Sónya Kovalévsky: Her Recollections of Childhood*. Trans. I. F. Hapgood. Century, New York, 1895.

4. *The New Oxford Annotated Bible with Apocrypha (NRSV)*. Oxford Univ. Press, New York, 2010.

5. *Vulgate New Testament*. Samuel Bagster and Sons, London, 1872.

6. S. Weil, *Waiting for God*. Trans. E. Craufurd. G. P. Putnam's Sons, New York, 1951.

7. ――――, *Gravity and Grace*. Trans. A. Wills. G. P. Putnam's Sons, New York, 1952.

How To Play Mathematics

MARGARET WERTHEIM

What does it mean to know mathematics? Since math is something we teach using textbooks that demand years of training to decipher, you might think the *sine qua non* is intelligence—usually "higher" levels of whatever we imagine that to be. At the very least, you might assume that knowing mathematics requires an ability to work with symbols and signs. But here's a conundrum suggesting that this line of reasoning might not be wholly adequate. Living in tropical coral reefs are species of sea slugs known as nudibranchs, adorned with flanges embodying hyperbolic geometry, an alternative to the Euclidean geometry that we learn about in school, and a form that, over hundreds of years, many great mathematical minds tried to prove impossible.

Sea slugs have at least the rudiments of brains; they generally possess a few thousand neurons, whose large size has made these animals a model organism for scientists studying basic neuronal functioning. This tiny number isn't nearly enough to enable the slug to formulate any representation of abstract signs, let alone an ability to mentally manipulate them, and yet, somehow, a nudibranch materializes in the fibers of its very being a form that genius-level human mathematicians didn't discover until the nineteenth century; and when they did, it nearly drove them mad. In this instance, complex brains were an impediment to understanding.

Nature's love affair with hyperbolic geometry dates to at least the Silurian age, more than 400 million years ago, when sea floors of the early Earth were covered in vast coral reefs. Many species of corals, then and now, also have hyperbolic structures, which we immediately recognize by the frills and crenellations of their forms. Although corals are animals, they have only simple nervous systems and can't be said to have a brain. A head of coral is actually a colonial organism made up of thousands of individual polyps growing together; collectively, they grow a vascular system, a respiratory system, and a crude gastrointestinal system through which all the individuals of the colony eat and breathe and share nutrients. Nothing like a brain exists, and yet the colony can organize itself into a mathematical surface disallowed by Euclid's axiom about parallel lines. Strike two against "higher intelligence."

Ask any high schooler what the angles of a triangle add up to, and she'll say, "180 degrees." That isn't true on a hyperbolic surface. Ask our student what's the circumference of a circle, and she'll say, "2π times the radius." That's also not true on a hyperbolic surface. Most of the geometric rules we're taught in school don't apply to hyperbolic surfaces, which is why mathematicians such as Carl Friedrich Gauss were so disturbed when finally forced to confront the logical validity of these forms, and hence their mathematical existence. So worried was Gauss by what he was discovering about hyperbolic geometry that he didn't publish his research on the subject: "I fear the howl of the Boetians if I make my work known," he confided to a friend in 1829. To their universal horror, other mathematicians soon converged on the same conclusion, and the genie of non-Euclidean geometry was let loose.

But can we say that sea slugs and corals *know* hyperbolic geometry? I want to argue here that in some sense they do. Absent the apparatus of rationalization and without the capacity to form mental representations, I'd like to postulate that these humble organisms are skilled geometers whose example has powerful resonances for what it means for us *humans* to know math—and also profound implications for teaching this legendarily abstruse field.

I'm not the first person to have considered the mathematical capacities of nonsentient things. Toward the end of Richard Feynman's life, the Nobel Prize–winning physicist is said to have become fascinated by the question of whether atoms are "thinking." Feynman was drawn to this deliberation by considering what electrons do as they orbit the nucleus of an atom. In the earliest days of atomic science, atoms were conceived as little solar systems, with the electrons orbiting in simple paths around their nuclei, much as a planet revolves around its sun. Yet in the 1920s, it became evident that something much more mathematically complex was going on; in fact, as an electron buzzes around its nucleus, the shape it makes is like a diffused cloud. The simplest electron clouds are spherical; others have dumbbell and toroidal shapes. The form of each cloud is described by what's called a Schrödinger equation, which gives you a map of where it's possible for the electron to be in space.

Schrödinger equations (named after the pioneering quantum theorist Erwin Schrödinger and his hypothetical cat) are so complicated that, when Feynman was alive, the best supercomputers could barely simulate even the simplest orbits. So how could a brainless electron be effortlessly doing what it was doing? Feynman wondered if an electron was calculating its Schrödinger equation. And what might it mean to say that a subatomic particle is calculating?

The world is full of mundane, meek, unconscious things materially embodying fiendishly complex pieces of mathematics. How can we make sense of this? I'd like to propose that sea slugs and electrons, and many other modest natural systems, are engaged in what we might call the performance of mathematics. Rather than *thinking* about math, they are *doing* it. In the fibers of their beings and the ongoing continuity of their growth and existence, they *enact* mathematical relationships and become mathematicians-by-practice. By looking at nature this way, we are led into a consideration of mathematics itself not through the lens of its representational power but instead as a kind of transaction. Rather than being a remote abstraction, mathematics can be conceived of as something more like music or dancing; an activity that takes place not so much in the writing down as in the playing out.

Music gives us a rich analogy by which to consider the idea of mathematics as performance, for you don't need to be able to write down music to be a musician—maybe if you want to play Mozart, but not in many other cases. Most folk music throughout history has been created by people who are sonically illiterate. Elvis Presley, Michael Jackson, Eric Clapton, and Jimi Hendrix all claimed not to read music. In a British TV interview, Paul McCartney said, "As long as the two of us know what we're doing, i.e., John and I, we know what chords we're playing and we remember the melody, we don't actually ever have the need to write it down or read it."

Indian classical music, easily as complex as the Western classical canon, is based on ragas that were generally transmitted aurally from master to student, not traditionally written down. In this millennia-old practice, music is recognized as an innately mathematical form: the

Sanskrit word *prastara* means the "study of mathematically arranging" ragas and rhythms into pleasing compositions. Ragas certainly can be written down (indeed, Indian musical notation dates back more than 2,000 years), and mathematics can be notated, but it doesn't *have* to be. There are lots of things *doing* math without a formal script, and I'd argue that it makes no sense to say that electrons or sound waves are following mathematical instructions any more than it makes sense to say that Jimi Hendrix was following a musical score. The possibility of writing down music is something apart from its performance, and math can be considered in a similar way. In short, the notation isn't the act.

Among my favorite mathematical performers are holograms, which enact a gorgeous operation called the Fourier transform. This extraordinarily complex, elegant equation is named in honor of Joseph Fourier, a mathematician and physicist who advised Napoleon and discovered what we now call the greenhouse effect (he called it the "hotbox" effect). The Fourier transform has been called the most useful piece of mathematics of all time; you rely on its power every time you make a cell phone call or listen to a piece of digitally recorded music. Music synthesis also results from clever applications of Fourier's equations. We'll get to the audio part in a moment, but first let's look at the visual face of this mathematical marvel.

Holograms differ from photographs in a fundamental way: a photo captures a two-dimensional rendering of light and shade and color, like a very detailed painting; meanwhile, when light shines through a holographic plate, it assembles into a three-dimensional replica of the original object, re-creating in light a simulacrum of that thing. The image you see with a hologram is sculptural, really occupying three-dimensional space, so you can move around and view it from different angles. Yet when you look at a holographic plate, there's no image at all, just a blur in which you may be able to discern speckled rings and dots. What's been captured on the plate is the Fourier transform of the object, which encodes more information and a different kind of information than a photo can.

Every object has a Fourier transform, and in theory we could calculate the transform of any object we desire and make a holographic plate to generate its form even though an actual physical object never existed. The emerging field of computer-generated holography (CGH) is trying to do just this. If it can be made to work, it will revolutionize computer games and animation; we'd be able to watch whole movies akin to the marvelous holographic projection of Princess Leia in the original *Star Wars* film.

Calculating transforms for complex objects requires vast computational powers and skills as yet unachieved by human CGH practitioners. Nonetheless, simple chemicals interacting with light on a piece of film manage to enact Fourier transforms of complicated scenes. Acting together, wave fronts of light and atoms execute a beautiful piece of mathematical encoding, and when the light plays back through the film they *do* the de-encoding. As such, where a photograph is a representation, a hologram is a performance.

Fourier came to his equation in the early 1800s, not to describe images (the origin of holograms dates to the 1940s), but to describe heat flow, and it turns out that his mathematics also leads to enormously powerful applications in the audio domain. Why does a piece by Mozart sound so different when played on a flute or a violin? One way of explaining it is that, although both instruments are playing the same sequence of notes, the Fourier transform of the sound produced by each one is different. The transform reveals the sonic DNA of the instrument's sound, giving us a precise description of its harmonic components (formally, it describes the set of pure sine waves that make up the sound). With software, audio engineers can analyze the transform of a musical recording and tell you what kind of instrument was playing; moreover, they can tweak the transform to bring out qualities they like and filter out ones they don't. By fiddling with the math, one can sculpt the sound to suit particular tastes.

Calculating Fourier transforms of sounds is a lot easier than calculating the transforms of visual scenes, and software engineers have created programs to simulate musical instruments (e.g., Apple's GarageBand), effectively giving users a sim-orchestra on their laptops for the price of an app. Advances in Fourier-based sound simulation have revolutionized the economics of the music business, including movie scoring. Now you don't need an actual orchestra to produce stirring strings to accompany a heroine's triumph; you can conjure them from the virtual depths, generated through mathematics.

Whereas music synthesis demonstrates how we can use mathematics to create something powerful out of a vacuum, here I'm more interested in what happens in actual concert halls. Each great hall has its own unique "sound"; each room acts as a filter for the music, tweaking and sculpting its Fourier transform. Contemporary acoustic engineers use Fourier techniques when designing new concert halls, manipulating the architecture of the space, for example, adding baffles in specific places, all aided by software that simulates how sounds will react within the space. If the engineers do their job well, there will be no "dead spots," and the hall will sing with warmth and resonance. Here we have a mathematical performance between the sound waves, the architecture, and the surfaces of the walls.

Some music schools now have electronic "practice rooms," where, through software, you can dial up a Fourier-based simulation of a cathedral or a tin shed and hear what your playing would sound like in different spaces. However, music connoisseurs will tell you that no simulation is a substitute for physical reality, which is why revered concert halls, such as Vienna's Musikverein, or New York's Carnegie Hall, won't be replaced by software any time soon. It's interesting that most of the best-rated halls were built before 1901, a fact that the acoustic legend Leo Beranek has attributed to their *lack* of fancy architecture (their resolutely shoe-box shape) and their lightly upholstered seats. From the perspective I'm adopting, even the *chairs* can be said to be participating in the mathematical performance enacted in a concert hall. Score another home run for nonsentience.

Since at least the time of Pythagoras and Plato, there's been a great deal of discussion in Western philosophy about how we can understand the fact that many physical systems have mathematical representations: the segmented arrangements in sunflowers, pinecones, and pineapples (Fibonacci numbers); the curve of nautilus shells, elephant tusks, and rams' horns (logarithmic spirals); music (harmonic ratios and Fourier transforms); atoms, stars, and galaxies, which all now have powerful mathematical descriptors; even the cosmos as a whole, now represented by the equations of general relativity. The physicist Eugene Wigner has termed this startling fact "the unreasonable effectiveness of mathematics." Why does the real world actualize math at all? And so much of it? Even arcane parts of mathematics, such as abstract algebras and obscure bits of topology, often turn out to be manifest somewhere in nature. Most physicists still explain this by some form of philosophical Platonism, which in its oldest form says that the universe is molded by mathematical relationships that precede the material world. To

Platonists, matter is literally *in-formed*, and guided by, a preexisting set of mathematical ideals.

In the Platonic way of seeing, matter (the stuff of everything) is rendered inert, stripped of power, and subordinated to ethereal mathematical laws. These laws are given ontological primacy, and matter is effectively a sideline to the "true reality" of the equations. Over the past half-century, this vision has been updated somewhat because now matter, or subatomic particles, have themselves been enfolded into the equations. Matter has been replaced by fields—as in electric and magnetic fields—and now it's the *fields* that follow the laws. Still, it's the laws that retain primacy and power, hence the obsession with finding an ultimate law, a so-called "theory of everything."

Platonism has always bothered me as a philosophy in part because it's a veiled form of theology—mathematics replaces God as the transcendent, *a priori* power—so if we want to articulate an alternative, we need new ways of interpreting mathematics itself that don't also slip into deistic modes. Thinking about math as performative points a way forward while also offering a powerful pedagogic model.

Corals and sea slugs construct hyperbolic surfaces, and it turns out that humans can also make these forms using iterative handicrafts such as knitting and crochet—you can *do* non-Euclidean geometry with your hands. To crochet a hyperbolic structure, one just increases stitches at a regular rate by following a simple algorithm: "Crochet *n* stitches, increase one, repeat ad infinitum." By adding stitches, you increase the amount of surface area in a regular way, visually moving from a flat or Euclidean plane into a ruffled formation that models the "hyperbolic plane." Mathematically speaking, the hyperbolic plane is the geometric opposite of the sphere: where the surface of a sphere curves toward itself at every point, a hyperbolic surface curves away from itself. We can define these different surfaces in terms of their curvature: a Euclidean plane has zero curvature (it's flat everywhere), a sphere has positive curvature, and a hyperbolic plane has negative curvature. In this sense, it is a geometric analog of a negative number (Figure 1).

Just as geometric relationships on a sphere are different to those on a flat plane—think of what you know about the surface of the Earth versus a flat piece of paper—so, they are different again on a hyperbolic surface. Whereas on a flat plane the angles of a triangle add up to

FIGURE 1. "Crochet Coral Reef" by Margaret and Christine Wertheim and the Institute For Figuring installed at the Smithsonian's National Museum of Natural History. Photo © Institute for Figuring. See also color image.

180 degrees, on a sphere they add up to more, and on a hyperbolic surface they add up to less. It's hard to appreciate this abstractly when you learn it from textbooks, as I did at university, but you can demonstrate it materially on a crocheted hyperbolic plane by stitching triangles onto the surface. You can also demonstrate visually that parallel lines diverge and other apparent absurdities. If Gauss had known how to crochet, he mightn't have been driven so bonkers.

It took a woman, the mathematician Daina Taimina at Cornell University, to discover hyperbolic crochet and to give mathematicians a tangible model of this form. I have conducted workshops about this with women all over the world, delighting in how much geometry can be conveyed through acts of making. There's also a link here with general relativity, because the discovery of the hyperbolic plane opened up a whole new era in geometric thinking, leading ultimately to generalized Riemannian geometry, which can describe any complexly curved surface, and is the mathematics underlying Albert Einstein's equations for the cosmos.

FIGURE 2. Mosaic tiling from the tomb of Hafez in Shiraz. Courtesy Wikipedia. See also color image.

Via handicrafts, we can introduce people to concepts about curved space time and multidimensional manifolds, leading with our fingers to questions about measuring the structure of the cosmic whole. We can see this as a form of "digital intelligence," and it's worth noting that iterated handicrafts (knitting, crochet, weaving) *were* the original digital technologies: their algorithmic "patterns" are literally written in code. It's no coincidence that computer punch cards were derived from the cards used in automated looms. Here, knowing emerges from hands performing mathematics: it is a kind of embodied figuring.

People talk about playing music, and mathematics can also be a form of play. One way of thinking about math is as a language of pattern and form, so when you play with patterns you are doing math. A beautiful example of mathematical pattern play can be seen with the great Islamic mosaicists who decorated mosques and palaces such as the Alhambra Palace in Granada in Spain with intricate tilings whose mathematical complexities are still a source of wonder (Figure 2).

Long before European geometers realized that there are only seventeen mathematically distinct tessellations of the plane—different ways of filling an area with a regular tiling pattern—medieval mosaicists working with their hands using the Hasba method knew about them all. Moreover, medieval Islamic tilers had also discovered aperiodic tiling, which is a way of filling a plane where the pattern never repeats. Western mathematicians discovered these tilings only in the 1960s, again after centuries of theorizing that such patterns were impossible. One of the magical qualities of aperiodic tilings is that they look simultaneously random and regular; as a geometric form of chaos, they are rule-based yet inherently unpredictable.

At first, when Western mathematicians (Sir Roger Penrose among them) discovered aperiodic tilings, these formations were thought to be just a mathematical curiosity; like hyperbolic surfaces, they seemed to defy common sense so that no one imagined such things could be present in the physical world (Figure 3). Prejudice was so intense that when the Israeli chemist Dan Shechtman announced in 1982 that he'd created a new type of crystal with an aperiodic structure, many fellow scientists refused to believe him. (Like Gauss, he too delayed publishing because of the supposedly absurd nature of his claims.) Shechtman's *quasicrystals* have brought about a paradigm shift in crystallography, in part because now we know that crystals can be chaotic, exhibiting order without repetition.

Lewis Carroll would have had a field day with this concept, which calls to mind the Red Queen's exhortation to Alice that, with practice, one can "believe six impossible things before breakfast." In 2009, after an intense search, a naturally occurring example of an aperiodic crystal was also found in the mineral icosahedrite (Figure 4). Strike three against intelligence as a prerequisite for *doing* mathematics.

As a nice coda to this story, in 2011 Shechtman was awarded the Nobel Prize in chemistry.

Proof that studying equations isn't the only path to mathematical insight also comes to us from Africa, where crafts people discovered fractals centuries ago. A wide variety of fractal patterns are incorporated into African textiles, hairstyling, metalwork, sculpture, painting, and architecture. One marvelous Ba-Ila village in southern Zambia is laid out in a fractal design reminiscent of the Mandelbrot set, that swirling icon of 1990s computer-graphic cool (Figures 5 and 6). In

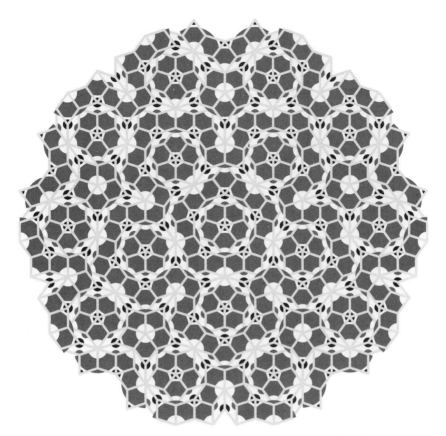

FIGURE 3. Aperiodic "Penrose" tiling pattern. Courtesy Wikipedia

his book *African Fractals: Modern Computing and Indigenous Design* (1999), mathematician Ron Eglash traces the story of the southern continent's priority in a branch of geometry that came into Western consciousness only around the turn of the twentieth century and didn't really flourish here until the development of computer graphics chips.

Sea slugs do math, electrons do math, minerals do math. Rainbows do an incredible mathematical performance, when you take into account the primary and secondary bows, the dark band between them, and the red and green arcs of light under the primary bow. Next time you see a good rainbow, stop and take a look at the space around it— there's so much going on; classical geometric optics doesn't begin to capture its complexity. A stunning piece of mathematical performance

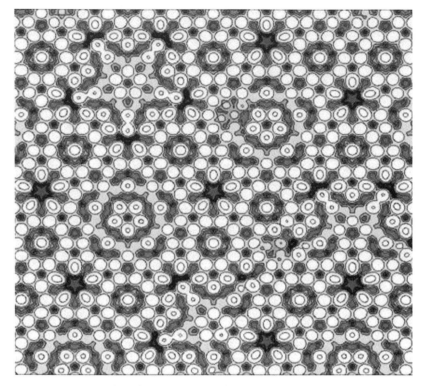

FIGURE 4. Image of an aluminum–palladium–manganese quasi-crystal surface. Courtesy Wikipedia. See also color image.

is enacted by a peregrine falcon as it hurtles toward its prey; with its head held straight so that it can fix one eye steadily on the quarry at a constant angle of 40 degrees, it swoops down at two hundred miles per hour in a perfect logarithmic spiral. Leonhard Euler's eighteenth-century formula, with its unique mathematical properties, is enacted here by a bird.

All around us, nature is playing mathematical games, and we too can join in the fun. Mathematics need not be taught as an abstraction; it can be approached also as an embodied practice, like learning a musical instrument. This notion doesn't invalidate what goes on in university classrooms or academic textbooks, since society needs professional mathematicians who can work with symbols, people such as Fourier and Bernhard Riemann, people who developed the math that

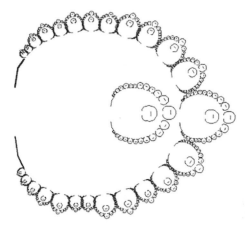

FIGURE 5. Fractal model for Ba-Ila village. From *African Fractals: Modern Computing and Indigenous Design* by Ron Eglash

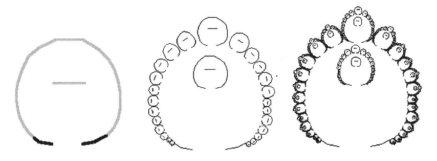

FIGURE 6. First three iterations of fractal model for Ba-Ila village. From *African Fractals: Modern Computing and Indigenous Design* by Ron Eglash

assists us to make cell phone calls, or determine the structure of the cosmos, and so much else besides. Because nature does *so much* mathematics, there will probably never be a time when professional "symbolizing" isn't profoundly useful. In 2016, the Nobel Prize in physics was awarded for "theoretical discoveries about topological phase transitions in matter"—astonishing, complex work that emerged out of the discovery of another kind of supposedly impossible object (the quasi-particle) and whose mathematical insights might pave the way for quantum computers.

By thinking about mathematics as performance, we liberate it from the straitjacket of abstraction into which it has been too narrowly confined. If you ask professional mathematicians what they love about their work, a likely answer is its beauty. "Euclid alone looked on beauty bare," wrote the poet Edna St. Vincent Millay in 1923, while the mathematician André Weil (brother of Simone) claimed that solving a hard mathematical problem topped sexual pleasure.

The professionals know that mathematics swings; they delight in its playfulness, the plasticity of its forms, and (after some initial shock) the absurdities it throws up. Hyperbolic surfaces, aperiodic tilings, Möbius strips, negative numbers, and zero all generated alarm at first, yet were ultimately embraced as gateways to new continents of mathematical wonder.

You don't have to be a symbol expert to appreciate this terrain. Just as humans are endowed with an ability to dance and play music (even if education too often crushes this out of us), so we have innate form-making and pattern-playing proclivities. Sea slugs, sound waves, and falcons do mathematics; Islamic mosaicists and African architects do it too. So can you.

Beauty Is Not All There Is to
Aesthetics in Mathematics

Robert S. D. Thomas

I draw attention to the discrepancy in philosophy of mathematics between the two main uses of terms involving "aesthetics" of which I am aware. It is a commonplace to admit or claim that "aesthetic considerations" influence choices in mathematical practice not only in pure mathematics but also in applied mathematics, where exclusively utilitarian considerations might be expected. On the other hand, discussions of aesthetics within philosophy of mathematics are concerned nearly exclusively with discussions of beauty.

In looking at aesthetic literature in philosophy of mathematics (e.g., [Plotnitsky, 1998]) in order to confirm that it mostly treats just beauty—however that is viewed—the closest that I have come to anyone's looking away from beauty is the essay by Nathalie Sinclair [2006] in the aesthetics volume that she coedited [Sinclair et al., 2006]. In their introductory chapter, the editors slip into writing of "the aesthetic feeling" [Sinclair and Pimm, 2006, p. 12] as though there were only one. That does not seem to be Sinclair's considered view, as she writes of what is "beautiful and interesting" [2006, p. 92] and acknowledges three different characteristics of "the aesthetic," the *motivational*, which attracts to what is not yet done and stimulates to do it, the *generative*, which guides mathematical moves that are not deductive, and the *evaluative*, which is the second-order appreciation of what has been done [2006, p. 89].

Having mentioned beauty, I should point out that I accept Rom Harré's view [1958] that judgments like that of beauty are "second-order" (his term) without necessarily wishing to accept his downgrading them to "quasi-aesthetic" (also his term). What I think Harré means by "second-order" is that, in order to view a proof, say, as elegant, his favorite such feature, one needs to know it first and to appreciate that its simplicity has been artfully achieved even if one does not know a clumsier proof. This seems to be Sinclair's evaluation, which she also calls "second-order."

The above discrepancy is an example of the tunnel vision to which all scholarship is prone.[1] Once things are published that confine aesthetics of mathematics to beauty in mathematics, there is a tendency—has been a tendency—to maintain that narrowness. This tendency is well established in English [Todd, 2008].[2] The resolution of the discrepancy is obvious; aesthetics in mathematics needs to consider more than just what is beautiful. Those discussing a number of other features can admit that those features are aesthetic. The special issue of *Philosophia Mathematica* [23, 2015, No. 2] on mathematical depth has no acknowledgment that its topic may sometimes be an aesthetic one, no indication of such a context, something one might expect even if it were argued about.

A reason why this is not done, which is a bad reason, is the low status of aesthetics in general within philosophy. This status is bemoaned even by its enthusiasts [Devereaux, 1998].[3] Mary Devereaux points out that "philosophers widely regard aesthetics as a marginal field." She continues,

> Aesthetics is marginal not only in the relatively benign sense that it lies at the edge, or border, of the discipline, but also in the additional, more troubling, sense that it is deemed philosophically unimportant. In this respect, aesthetics contrasts with areas like the philosophy of mathematics, a field which, while marginal in the first sense, is widely regarded as philosophically important.

To be at the margin of the margin is to risk falling off the edge. (The cure for that is to be edgier.) Whatever may be the case within philosophy in general, if one is to consider mathematical practice, which is performed by humans with values, their mathematical choices must come into play, and those choices are based on aesthetic considerations, among others. Aesthetics in mathematical practice is not marginal in either sense.

My argument is for two theses. "Interesting" is an aesthetic feature seen in all published mathematics. This phenomenon may be interesting, but I do not see it as important in itself. It implies my second thesis, though, that beauty is not all there is to aesthetics in mathematics any more than beauty is all there is to aesthetics outside of mathematics. This I do not see as interesting, but the default contrary view is importantly wrong (not to mention the further limitation to proofs).

1. Being Interesting Is an Aesthetic Value

In order to claim that attribution of aesthetic considerations unrelated to beauty is not just eccentric because aesthetic considerations *have* to relate to beauty, I offer some second-hand historical justification. The book, *The Future of Aesthetics* by Francis Sparshott [1998] actually says enough about the past of aesthetics for present purposes. The term "aesthetik," he says, was coined by Alexander Gottlieb Baumgarten [1735] to cover the nonlogical side of philosophical considerations, matters of more and less rather than yes and no. This is plainly the sense in which what are still called "aesthetic considerations" influence mathematical practice. How it has survived the narrowing of aesthetics to beauty and the arts already in the eighteenth century, I do not know. Some word has to cover nonlogical matters of degree, and "aesthetics" has remained available. "Valid" of a proof is an evaluative term indicating a position on a Boolean scale. Aesthetic evaluations are those that are not logical but have to do with a scale of merit. "Short" of a proof is descriptive and not logical, but it is not in itself an aesthetic judgment. To say, however, that one proof is shorter than another, *ceteris paribus*, could be an aesthetic evaluation, if only a minor one. How minor would depend on the lengths compared. Almost any conceptual proof of the four-color theorem would be shorter and aesthetically preferable to the computer-assisted proof, while being preferable in other respects as well.

I claim that, as well as "beautiful" and sometimes "deep," "interesting" is an important aesthetic category—indeed that "interesting" is a *sine qua non* of publishable mathematical research. The main basis for this claim is my ten years of experience as managing editor of a mathematics journal. For the whole of that time, the criteria that I asked referees to use in recommending acceptance of a manuscript were whether it was original, correct, and interesting. One does not want to publish what has already been published or what is wrong or what is new and correct but of no interest. Arithmetic alone supplies an infinite sequence of such results, since there are always larger and larger number pairs that have not been added. If this were research, M.Sc. theses could do additions and Ph.D. theses subtractions. These sums and differences must be rejected on the ground of lack of interest, since they cannot be rejected for being either wrong or well-known. While these particular results lack interest because the method of *producing* them is well-known, I say without fear of contradiction that there are other ways to lack interest. Writers create documents of some originality that are mathematical but lack mathematical interest; sometimes they submit them to a philosophy journal.

Perhaps being interesting, in spite of being a matter of degree, is merely an epistemic condition; that would get it out of the aesthetic box. It is partly but not just epistemic, because the uninteresting sums do tell their reader something that the reader did not already know, and the mathematical results that their authors think are of philosophical though not mathematical interest do not follow well-trodden paths. They do tell a reader something that the reader definitely does not know but probably does not want to know. Not wanting to know something is, it seems to me, a negative aesthetic judgment.[4] One can hardly deny an epistemic component to mathematical interest, but different mathematical results *feel* different. Learning mathematics does not necessarily make one feel interested, as generations of school leavers attest. Being interesting is not just an epistemic condition, but is it associated with other aesthetic judgments? As soon as I formulated this question in the spring of 2015, evidence came unbidden.

In the then current issue of *Philosophia Mathematica*, the paper on mathematical beauty [Inglis and Aberdein, 2015] has the second sentence, "Mathematicians talk of 'beautiful', 'deep', 'insightful', and 'interesting' proofs, and award each other prizes on the basis of these assessments." Despite the fact that being interested is very much a matter of how one feels, "interesting" does not occur again in the paper. In particular, it is not one of the eighty adjectives compared with "beautiful." I do not disagree with the authors that interest seems orthogonal to beauty. But they clearly regard it as aesthetic anyway.

In the then current issue of *The Mathematical Intelligencer*, being interesting is singled out in the cooking column [Henle, 2015] as an aesthetic quality common to mathematics and wine. In both cases, one needs some level of sophistication to find interest rather than just learning something or quenching thirst. Having made this point about wine, mathematician Jim Henle continues,

> Mathematics is much the same. It's more than useful; it's engaging. The fact that two plus two is four satisfies a primitive need, but a complex mathematical structure holds our interest. Mathematical

ideas are enigmatic and charming. They yield treasures and they keep secrets. Mathematical structures appear different in different contexts. Local changes force global transformations. Mathematics entertains us and we treasure its mysteries.

The same, but also different. One difference is that interest for mathematics is essential, beauty an option; for wine, a pleasing flavor is essential with interest an option. The distinguished aesthetician Frank Sibley devoted a long time to a substantial essay [2002a], which he died before publishing, on the aesthetic value of tastes and smells. Jim Henle did not make up either side of his analogy.

There are two verbal matters to do with "interest" that need to be mentioned. Aesthetic judgments, like other judgments that are meant to have objectivity or intersubjectivity, are supposed to be disinterested. How then can a judgment of interest be disinterested? Because the sense of "interest" that disinterest avoids means dependent on the thought of or "desire for the use or possession of their objects" ([Cooper, 1711], as quoted in the *Stanford Encyclopedia of Philosophy* article "18th century British aesthetics" of 2014). Accordingly, "disinterested interest" is no oxymoron, only a limitation. I return to disinterest.

The other verbal matter is the complementarity of ways, independent of those of the previous paragraph, in which we use "interest." As I have often told students of a compulsory course that they would not have chosen, one can be passively the prisoner of what interests one willy nilly, as everyone knows, or one can deliberately *take an interest* in something that one decides to pay attention to, like a course that one is taking, for example. The latter is an act of will that is carried out by reading attentively and perhaps doing exercises, but surprisingly often (always in my experience) it leads to becoming reactively interested by the material. An example is my recent experience of the *Spherics* of Theodosios [Ver Eecke, 1959]. This second-century BCE treatise on circles on a sphere, although it held its place in the *quadrivium* as long as that lengthy tradition lasted, has been little thought of since. It did not reactively interest Thomas Heath in particular [1921, vol. 2, pp. 248–252], and he did not take an active interest in it. A few years ago, I took sufficient interest in it to translate it from Greek.[5] This work was definitely an act of will; my Greek is not good enough for any document to entice me to study it. But when I became familiar with it,

I found that it had mathematical interest enough to want to pass it on. As a result, I have written something I call an appreciation of the first of its three books [Thomas, 2018]. The more one knows about something valuable and mathematical, the more interest one finds in it. Mathematics that is not of sufficiently *general* interest is published in specialized journals where readers know enough about the subject to be reactively interested or to take an active interest.

The sense of "interest" at issue can perhaps be conveyed more clearly in terms of attention. It is an important feature of advertising and journalism to attract attention, to interest a person in the reactive sense. Good journalism and much other writing attempts to hold attention, to develop the reader's interest after initially grabbing it. This is our topic here, although since, being reactive, it must be begun; a bit of grabbing is part of the package. "It is a truth universally acknowledged, that a single man in possession of a good fortune must be in want of a wife."[6] Interest in anything whatever in the active sense is always possible; one can pay attention to whatever one wants to pay attention to. And finally, the kind of interest to which attention is irrelevant is the kind referred to in the phrase "one's own best interests." Attention may advance those interests, but it does not help to define them. This is the interest avoided in disinterest.

In these terms, it is reactive interest that indicates the aesthetic value. Disinterested aesthetic contemplation is nontrivial to describe. Sibley quotes [2002a, p. 230] with approval this description:

> Thus we may define an interest in an object X for its own sake as a desire to go on hearing, looking at, or in some other way having experience of X, where there is no reason for this desire in terms of any other desire or appetite that the experience of X may fulfil, and where the desire arises out of, and is accompanied by, the thought of X . . . I shall respond to the question "Why are you interested in X?" . . . with a description of X. [Scruton, 1974, p. 148]

I believe that the sense of "interested in" in the question is "paying attention to." Scruton is not arguing, with me, that being interesting is itself an aesthetic value. When giving a description of X to explain why something is of aesthetic value, one frequently mentions nonaesthetic characteristics. This is true also of why something mathematical is interesting.[7]

Scruton's description illustrates that aesthetics cannot have firm boundaries. This is a fact (no necessary or sufficient conditions) emphasized by Sibley [1959], "Accordingly, when a word or expression is such that taste or perceptiveness is required in order to apply it, I shall call it an aesthetic term or expression, and I shall, correspondingly, speak of aesthetic concepts or taste concepts." I am claiming that the aesthetic has been drawn much too narrowly in discussion of mathematics. What is interesting in mathematics and *how* interesting (i.e., more or less rather than in what way) are very much matters of trained taste.

"Interesting" is not the same thing as the more objective property "important," which is not so subject to disagreement. The taste of wine is not in any absolute sense important, however interesting it may be, but it is of enormous commercial importance. Likewise, a piece of mathematics can be of importance either for application outside mathematics or for mathematical use, but it can be interesting without having either of those nonaesthetic values. Unfortunately, importance is paid more attention than interest in mathematical education. Since a most important aspect of teaching is the engagement of students' interest, this is a mistake. The good effects of Martin Gardner on young readers, now alas all grown up, are widely reported by them. Gardner, in his column in the *Scientific American* from 1956 to 1981, chiefly revealed the interest inherent in the topics he chose to expound. No one suggests that he manufactured that interest. He had a nearly unique ability to find and expose it and in consequence interested thousands of persons who became mathematicians and many more others. Education needs elements of this skill, as Gardner himself maintained [Mulcahy and Richards, 2014]. Discussion on what is interesting (and how) is probably a necessary preliminary. This has not been being done. An exception is Wells [2015, Ch. 3–6].

As I am not concerned here with beauty, I merely remark that mathematical interest is not infected with the difficulties that feminists can and do find in male-gendered considerations of beauty since Plato's *Symposium* [Sparshott, 1998, p. 15].

I have set out the fact that being interesting is regarded by some as an aesthetic value; it seems to me that such evidence is more important than argument. But as it happens there is an argument available, for what it is worth in such a context. It is rightly said that aesthetic

considerations weigh with mathematical researchers when doing research and not just afterward. I claim that those considerations, while
they may sometimes have to do with beauty, more frequently have to do
with judgments of what will be interesting. That is to say that research
is driven more by curiosity-based first-order judgments than second-
order judgments, which can only be made of results. It is incoherent to
claim that in searching for one knows not what (otherwise it would not
be research) one is strongly influenced by the appearance of what is to
be found. One could easily be influenced by what one *hopes* to find, and
no doubt often is, but that is as much about its interest as its potential
beauty. Conjectures are important, but they are not starting points.
One way that hopes can work is to motivate adjusting premises to allow
preferable results or proofs.[8]

It would be courteous to remark that Sinclair's "motivational" and
"generative" characteristics of the aesthetic are close to "being interesting," since being interesting applies to what one has not yet done and I
think also to what one has not yet read. Even reading mathematics requires some motivation. Surely one cannot appreciate as beautiful what
one has not yet done (and may not do) or what one has not yet read? In
both cases, one can be interested and often is.

I have no desire to put down beauty, only to elevate interest from
invisibility to its place of importance. There are different ways to be
interesting, but this is not the place for a catalog. One way currently
discussed is explanatoriness [Hafner and Mancosu, 2005; Tappenden,
2008; Baker, 2017]. It is regarded as a value of proofs almost universally,
although like beauty [Rota, 1997], it can be put down [Zelcer, 2013].
Proofs can be explanatory (or not [Resnik and Kushner, 1987][9]) of what
they prove, but also results can explain other results like the interval of
convergence of the series for $1/(1 + x^2)$. Much philosophical discussion
of explanation extends from mathematics to physical *explananda* [Baker,
2005], but that need not concern us here. Within mathematics various
ways to be explanatory are identified in Hafner and Mancosu [2005,
Section 3], where distinct ways of being explanatory are put forward,
the evidence coming down to the fact that mathematicians find them
to be explanatory to some degree. All of their analysis is about just one
way of being interesting. That there is so much room for differences of
opinion is evidence that being explanatory and interesting more generally are aesthetic qualities.

It must be admitted that, as the use of "interesting" in fiction shows, it is not the highest aesthetic value there. Crime fiction, fantasy, romances, and science fiction are often page-turners without being held to be of great literary value. It is unlikely that it is the highest praise for mathematics either. Both beauty and depth are more highly valued and less time-dependent [Wells, 1988, 1990]; all I am saying is that being interesting is necessary. The booklet *A Manual for Authors of Mathematical Papers* [AMS, 1990] warns not to try to publish detail that it is good to work out "since it is likely to be long and un-interesting" (quoted by Sinclair [2011]).

It is no part of my claim for interest that only mathematics is interesting. I have no idea even whether it is uniquely important in mathematics. In history, literary criticism, or any other discipline, what is written has also to be interesting to be publishable. While being written interestingly is a positive feature no doubt, the bad reputation of academic writing in general suggests that the interest needed in other subjects is in the subject matter for reasons to do with that subject matter. A history essay is of more interest as the events described are of more importance. Literary criticism is of more interest as the literature discussed is better. Economics is of more interest as the phenomena explained are more widespread or important in some other way. Mathematics can be of interest for this sort of reason too. There is limited interest for its own sake in the unsolvability in integers of Pythagoras-like relations for higher powers than two, but the proof of Fermat's famous conjecture was of great interest because centuries of effort had rendered it important beyond its raw material. As in many other examples of important mathematical accomplishments, active interest was taken on account of the importance. I do not mean to suggest that this active kind of interest is aesthetic, but that the reactive sort is. It is also the more frequent motivation—especially in pure mathematics.

2. Conclusion

I conclude with a pair of contrasting analogies. Much mathematical effort is more like landscape gardening than like picture drawing. I take picture drawing to begin with a blank sheet on which the artist represents something imagined or seen, a chief aim being to create something of value. The artist is free, because the page is blank, in the choice of what is to be represented, which need not be something seen, and in how it is represented. Mathematical creation is not so free, hence the contrasting analogy of the landscape gardener, who needs a good grasp of the topography before getting down to creating something beautiful, which needs to be based on that topography. When H.S.M. Coxeter handed out copies of the preliminary edition of his book *Projective Geometry* [1974] to his undergraduate students in 1963, the preface included a sentence to the effect that the only mention of cross ratio in the book was in that sentence.[10] He drew our attention to this, regarding it as aesthetically pleasing to avoid all use of cross ratio in his landscaped development of elementary projective geometry. He knew the terrain well and was able to accomplish this aim because of his mastery of it. But Desargues had been dead for three hundred years[11]; much projective geometry had been done with the aim of understanding the topography, curiosity-driven research that had the aim of finding interesting projective properties, initially in Euclidean space and certainly not avoiding cross ratio. It is in the aesthetic space I have been writing about that the complementary scientific pursuit of what is interesting and artistic pursuit of what is beautiful interact. A lot of mathematics does not get past mapping the topography. G. H. Hardy is often quoted as writing "Beauty is the first test: there is no permanent place in the world for ugly mathematics" [1967, p. 85]. Being interesting is apparently test zero, because without being interesting it is not even ugly mathematical research. *Eventually* one wants everything to be beautiful and so to be permanent, but these things can take time. The problematic status of the parallel postulate was regarded as an aesthetic blemish on geometry for more than two thousand years before it was eliminated by clarifying that there is more to geometry than the *Elements*.[12] *Ars longa; vita brevis.*

Acknowledgments

I am grateful for comments on earlier versions of this to Carlo Cellucci, Jim Henle, Peter Lamarque, Colin McLarty, Marcus Rossberg, Nathalie Sinclair, Jean Paul Van Bendegem, David Wells, and Christian Wenzel.

Notes

1. An example of this must come from another topic. In 1898, H. B. Swete published an important commentary on the Gospel according to Mark, in which he accepted what

evidence there was that the book was written in a particular place for readers *in that place*. For a century, the assumption that each of the four Christian gospels was written for a separate community increasingly dominated scholarship, leading to creative and sometimes fanciful constructions of these supposed separate communities, about which nothing was known. This scholarly consensus was first challenged by Bauckham [1998], on which the above sketch depends. Contradictorily, during the same century there arose the even more dominant view that two of the Gospels depend on that according to Mark, which had obviously circulated to wherever they were written, and the famous source Q, which was also accessible to both writers. I return to this idea in note 12.

2. Most aesthetic literature on mathematics does not refer to anything before Harré [1958] except the book by Hardy [1967]. The situation is somewhat different in French; see, e.g., Sinclair [2011], which cites Hermite and Stieltjes [1905] (negative), Poincaré [1908] (heuristic), Hadamard [1945] (psychological), and Le Lionnais [1948] (taxonomic).

3. I owe reference to Devereaux to Robert Kraut [2007].

4. It is perhaps clearer that the decision that a *question* is uninteresting is nonepistemic since in the question there is no knowledge to call forth an epistemic judgment.

5. It had already been translated from a Latin translation in the eighteenth century, but I did not know that.

6. The famous first sentence of *Pride and Prejudice* by Jane Austen (1775–1817).

7. Sibley wrote a whole essay [1965] on the mysterious relation in the arts between aesthetic and nonaesthetic properties, in particular that, as Scruton says, one's explanation of aesthetic value is descriptive, and, as Sibley says, that description involves mainly nonaesthetic properties, which somehow add up to something aesthetic. A mathematical example of this is that, in explaining why a proof is interesting, one might invoke the so-called purity of its method.

8. Rota [1997, p. 178], quoted in Cellucci [2014].

9. The negativity of Resnik and Kushner seems to be based on their notion that explanatoriness is a matter of yes and no rather than of degree, which it is as an aesthetic feature. They quote Davis and Hersh [1982, p. 299] before explanatoriness became a common concern as writing that the prime-factorization proof of the irrationality of $\sqrt{2}$ "exhibits a higher degree of aesthetic delight" than the Pythagorean proof because it "seems to reveal the heart of the matter." What they clearly regarded as a matter of degree, since they use the word, I venture would later have been termed "explanatoriness."

10. Disappointingly, the sentence was soon changed to, "In particular, the only mention of *cross ratio* is in three exercises at the end of Section 12.3."

11. Girard Desargues (1591–1661) was a founder of projective geometry.

12. Just as there is more to aesthetics than beauty. This is a further example of what was discussed in note 1. It could have been noticed at any time after the writing of the first book of Spherics (almost certainly before Euclid, who uses results from what is now the second and third books) that the postulates other than the parallel postulate are satisfied by points and great circles on a sphere, but attention was reserved for the intended model.

References

American Mathematical Society [1990]: *A Manual for Authors of Mathematical Papers*. Providence, RI.

Baker, Alan [2005]: "Are there genuine mathematical explanations of physical phenomena?" *Mind* **114**, 223–238.

Baker, Alan [2017]: "Mathematics and explanatory generality," *Philosophia Mathematica (3)* **25**, 194–209.

Bauckham, Richard [1998]: *The Gospels for All Christians*. Grand Rapids, MI, and Cambridge, U.K.: Eerdmans.

Baumgarten, Alexander Gottlieb [1735]: *Meditationes philosophicae de nonnullis ad poema pertinentibus*. Halle im Magdeburgischen: Grunert. *Reflections on Poetry*. K. Aschenbrenner and W. B. Hoelther, trans. Berkeley: University of California Press, 1954.

Cellucci, Carlo [2014]: "Mathematical beauty, understanding, and discovery," *Foundations of Science* **20**, 339–355.

Cooper, A. (Third Earl of Shaftesbury) [1711]: *Characteristics of Men, Manners, Opinions, Times*. Indianapolis: Liberty Fund, 2001.

Coxeter, H.S.M. [1974]: *Projective Geometry*. 2nd ed. Berlin: Springer.

Davis, P. J., and R. Hersh [1982]: *The Mathematical Experience*. Boston: Houghton Mifflin.

Devereaux, Mary [1998]: "The philosophical status of aesthetics" at http://aesthetics-online .org/?page=DevereauxStatus. Accessed July 2016.

Hadamard, J. [1945]: *The Mathematician's Mind: The Psychology of Invention in the Mathematical Field*. Princeton, NJ: Princeton University Press, 1945. Reprinted New York: Dover, 1954.

Hafner, J., and P. Mancosu [2005]: "The varieties of mathematical explanation" in P. Mancosu, K. Jørgensen, and S. Pedersen, eds., *Visualization Explanation and Reasoning Styles in Mathematics*, pp. 215–250. Berlin: Springer.

Hardy, G. H. [1967]: *A Mathematician's Apology*. Cambridge, U.K.: Cambridge University Press. 1st ed., 1940.

Harré, Rom [1958]: "Quasi-aesthetic appraisals," *Philosophy* **33**, 132–137.

Heath, Thomas [1921]: *A History of Greek Mathematics*. 2 vols. Oxford, U.K.: Clarendon Press.

Henle, Jim [2015]: "The wine column," *The Mathematical Intelligencer* **37**, No. 1, 86–88.

Hermite, Charles, and T. J. Stieltjes [1905]: *Correspondance d'Hermite et de Stieltjes*. 2 vols. B. Baillaud and H. Bourget, eds. Paris: Gauthier-Villars.

Inglis, Matthew, and Andrew Aberdein [2015] "Beauty is not simplicity: An analysis of mathematicians' proof appraisals," *Philosophia Mathematica (3)* **23**, 87–109.

Kraut, Robert [2007]: *Artworld Metaphysics*. Oxford, U.K.: Oxford University Press.

Le Lionnais, François [1948]: "La beauté en mathématiques," in [Le Lionnais, 1962], pp. 437–465. Translated as [Le Lionnais, 1971a].

——, ed. [1962]: *Les Grands Courants de la Pensée Mathématique*. 2nd ed. Paris: Blanchard. 1st ed. 1948.

—— [1971a]: "Beauty in mathematics", in [Le Lionnais, 1971b], pp. 121–158.

——, ed. [1971b]: *Great Currents of Mathematical Thought*. Vol. 2. *Mathematics in the Arts and Sciences*. Trans. C. Pinter and H. Kline from the second edition. New York: Dover.

Mulcahy, Colm, and Dana Richards [2014]: "Let the games continue," *Scientific American* **311**, No. 4, 90–95.

Plotnitsky, Arkady [1998]: "Mathematics and aesthetics" in *Encyclopedia of Aesthetics*, Vol. 3, pp. 191–198. Michael Kelly, ed. Oxford, U.K.: Oxford University Press.

Poincaré, H. [1908]: "Mathematial creation" in James R. Newman, ed., *The World of Mathematics*, Vol. 4, pp. 2041–2050. London: Allen and Unwin and New York: Simon and Schuster, 1956. From H. Poincaré, *Foundations of Science: Science and Hypothesis, the Value of Science, Science and Method*. George Bruce Halsted, trans. New York: Science Press, 1913.

Resnik, Michael, and David Kushner [1987]: "Explanation, independence, and realism in mathematics," *British Journal for the Philosophy of Science* **38**, 141–158.

Rota, G.-C. [1997]: "The phenomenology of mathematical beauty," *Synthese* **111**, 171–182.

Scruton, Roger [1974]: *Art and Imagination: A Study in the Philosophy of Mind*. London: Methuen.

Sibley, Frank [1959]: "Aesthetic concepts," *The Philosophical Review* **68**, 421–450. Revised reprint in J. Margolis, ed., *Philosophy Looks at the Arts*. New York: Scribners, 1962, and in [Sibley, 2002b] as Chapter 1, pp. 1–23.

—— [1965]: "Aesthetic and nonaesthetic," *The Philosophical Review* **74** (1965), 135–159. And in [Sibley, 2002b] as Chapter 3, pp. 33–51.

—— [2002a]: "Tastes, smells, and aesthetics," posthumously published in [Sibley, 2002b], as Chapter 15, pp. 207–255.

—— [2002b]: *Approach to Aesthetics: Collected Papers on Philosophical Aesthetics*. John Benson, Betty Redfern, and Jeremy Roxbee Cox, eds. Oxford, U.K.: Clarendon Press, 2002.

Sinclair, Nathalie [2006]: "The aesthetic sensibilities of mathematicians" in [Sinclair et al., 2006], pp. 87–104.

—— [2011]: "Aesthetic considerations in mathematics," *Journal of Humanistic Mathematics* **1**, 2–31.

Sinclair, Nathalie, and David Pimm [2006]: "A historical gaze at the mathematical aesthetic" in [Sinclair et al., 2006], pp. 1–17.

Sinclair, Nathalie, David Pimm, and William Higginson, eds. [2006]: *Mathematics and the Aesthetic: New Approaches to an Ancient Affinity*. Berlin: Springer.

Sparshott, Francis [1998]: *The Future of Aesthetics*. University of Toronto Press.

Swete, H. B. [1898]: *The Gospel According to St Mark: The Greek Text with Introduction, Notes, and Indices*. London: Macmillan.

Tappenden, Jamie [2008]: "Mathematical concepts and definitions" in Paolo Mancosu, ed., *The Philosophy of Mathematical Practice*, pp. 256–275. Oxford, U.K.: Oxford University Press.

Thomas, R.S.D. [2018]: "An appreciation of the first book of Spherics," *Mathematics Magazine* **91**, 3–15.

Todd, Cain S. [2008]: "Unmasking the truth beneath the beauty: Why the supposed aesthetic judgements made in science may not be aesthetic at all," *International Studies in the Philosophy of Science* **22**, 61–79.

Ver Eecke, P. [1959]: *Les Sphériques de Théodose de Tripoli*, Paris: Blanchard.

Wells, David [1988]: "Which Is the most beautiful?" *The Mathematical Intelligencer* **10**, No. 4, 30–31.

—— [1990]: "Are these the most beautiful?" *The Mathematical Intelligencer* **12**, No. 3, 37–41.

—— [2015]: *Motivating Mathematics: Engaging Teachers and Engaged Students*. Singapore: Imperial College Press.

Zelcer, Mark [2013]: "Against mathematical explanation," *Journal for General Philosophy of Science* **44**, 173–192.

The Science of Brute Force

MARIJN J. H. HEULE AND OLIVER KULLMANN

Recent progress in automated reasoning and supercomputing gives rise to a new era of brute force. The game changer is "SAT," a disruptive, brute reasoning technology in industry and science. We illustrate its strength and potential via the proof of the Boolean Pythagorean triples problem, a long-standing open problem in Ramsey theory. This 200-terabyte proof has been constructed completely automatically—paradoxically, in an ingenious way. We welcome these bold new proofs emerging on the horizon, beyond human understanding—both mathematics and industry need them.

Many relevant search problems, from artificial intelligence to combinatorics, explore large search spaces to determine the presence or absence of a certain object. These problems are hard because of combinatorial explosion, and they have traditionally been called infeasible. The brute force method, which at least implicitly explores all possibilities, is a general approach to systematically search through such spaces.

Brute force has long been regarded as suitable only for simple problems. This notion has changed in the past two decades because of the progress in satisfiability (SAT) solving, which by adding brute reason renders brute force into a powerful approach to deal with many problems easily and automatically. Search spaces with far more possibilities than the number of particles in the universe may be completely explored.

SAT solving determines whether a formula in propositional logic has a solution, and its brute reasoning acts in a blind and uninformed way—as a feature, not a bug. We focus on applying SAT to mathematics as a systematic development of the traditional method of proof by exhaustion.

Can we trust the result of running complicated algorithms on many machines for a long time? The strongest solution is to provide a proof, which is also needed to show correctness of highly complex systems;

highly complex systems are everywhere, from finance to health care to aviation.

Many problems arising from areas such as Ramsey theory and formal methods appear to be intrinsically hard and may be only solvable by SAT. Any proof for such problems may be huge, in which case mathematicians will not be able to produce a paper proof. The enormous size of such proofs hardly influences confidence in the correctness, as highly trusted systems can validate them.

We argue that obtaining such results is meaningful regardless of our ability to understand them.

The Rise of Brute Force

We all know that brute force does not work, or at least is brutish, do we not? In our case, it is even "brute reasoning."

I can stand brute force, but brute reason is quite unbearable. There is something unfair about its use. It is hitting below the intellect.

—O. Wilde

A mathematician using "brute force" is a kind of barbaric monster, is she not? Case distinctions play an important role for thinking, but if the number of cases gets too big, it seems impossible to obtain an overview, and one has to slavishly follow the details. But perhaps this is what our times demand?

In the beginning of the twentieth century, there was a very optimistic outlook for mathematics. Gödel's incompleteness theorem seemed to destroy the positive spirit of the time, famously expressed by Hilbert's "We must know. We will know." That said, even Gödel anticipated the relevance of SAT solving in his letter to von Neumann[1], shifting the attention to finitizing infinite problems. Today, SAT solving on high-performance computing systems enables us to conquer problems of high complexity, driven by practice. This combination of enormous computational power with "magical brute force" can now solve very hard combinatorial problems, as well as proving safety of systems such as railways.

Our guiding example is the *Pythagorean triples problem* [17, 27], a typical problem from Ramsey theory: We consider all partitions of the set

$\{1, 2, \ldots\}$ of natural numbers into finitely many parts, and the question is whether always at least one part contains a Pythagorean triple (a, b, c) with $a^2 + b^2 = c^2$. For example, when splitting into odd and even numbers, then the odd part does not contain a Pythagorean triple (because of the rule that states odd plus odd = even), but the even part contains, for example, $6^2 + 8^2 = 10^2$. We show that the answer is *yes* [17], when partitioning into two parts, and we conjecture the answer to be *yes* for any finite size of the partition.

To solve the *Boolean Pythagorean triples problem*, it suffices to show the existence of a subset of the natural numbers, such that any partition of that subset into two parts has one part containing a Pythagorean triple. We focus on subsets $\{1, \ldots, n\}$, and determined by SAT solving that the smallest n for which the property holds is 7,825. Plain brute force cannot help, since 2^{7825}, the number of possible partitions into two parts, is way too big. So really "clever" algorithms are needed. An interesting aspect here is that there is no known ordinary mathematical existence proof for any form of the Pythagorean triples problem, even when generalizing the problem from triples $a^2 + b^2 = c^2$ to tuples $t_1^2 + \cdots + t_{k-1}^2 = t_k^2$. Only computational proofs are known, and, so far at least, only SAT solving can deal with the harder problems. We show that $\{1, \ldots, 10^7\}$ can be partitioned into *three* parts, such that *no part* contains a Pythagorean triple. Thus, if there is an n such that every 3-partitioning of $\{1, \ldots, n\}$ has a part containing a Pythagorean triple, then $n > 10^7$. Because of this enormous size, it is thus conceivable that the truth of the *three-valued Pythagorean triples problem* might never be known.

Before considering the solution process, one may ask, why should we care? Are there problems for which such reasoning is really useful? Yes, the same techniques are used to prove correctness of hardware and software systems. Finding a bug in a large hardware system is essentially the same as finding a counterexample, and thus is similar to finding a partition avoiding all Pythagorean triples. *Proving* correctness of a system, that is, that there is no counterexample, is similar to proving that each partition must contain some Pythagorean triple. SAT solving has revolutionized hardware verification [6], and now SAT can come to the rescue of mathematics, solving very hard combinatorial problems previously completely out of reach. This collaboration works in both directions, as the applications in mathematics, especially Ramsey theory, sharpen SAT algorithms: the cube-and-conquer method [18] was developed for computing van der Waerden numbers [1], and recently the cube-and-conquer solver Treengeling[2] won the parallel track of the 2016 SAT competition.[3] Deeper mathematical investigations into the structure of the SAT instances could help with understanding and improving SAT in general.

Well-known early mathematical proofs using *proof by exhaustion* are the four-color theorem [39] and the proof that no projective plane of order 10 exists [26]. Given a set of variables with finitely many values, a case-split explores all possible values. The former is actually a rather small case-split by modern standards (only hundreds of cases). The latter invokes a larger, but also man-made case-split (billions of cases), for which it can be determined in advance whether this will succeed. In contrast, we have currently no way of knowing whether the SAT solver's "magic" is sufficient to solve a given problem.

Throughout this article, we use the *Boolean Schur triple problem* as an example: Does there exist a red/blue coloring of the numbers 1 to n, such that there is no monochromatic solution of $a + b = c$ with $a < b < c \leq n$. Compared to the Boolean Pythagorean triples problem, all natural numbers are involved, not just square numbers. As a result, there are many more triples, and unsatisfiability is reached much sooner. For $n = 8$, such a coloring exists: color the numbers 1, 2, 4, and 8 red and 3, 5, 6, and 7 blue. However, such a coloring is not possible for $n = 9$. A naive brute force algorithm would consider all $2^9 = 512$ possible red/blue colorings. We will show that with brute reasoning only six (or even four) red/blue colorings need to be evaluated.

The Art of SAT Solving

A SAT problem uses Boolean variables v (they can be assigned to either `true` or `false`), which are constrained using clauses, which are disjunctions of literals x. Literals are either variables $x = v$ or their negations $x = \bar{v}$. A literal x (or $\bar{x}$) is `true` if the corresponding variable v is assigned to `true` (or `false`, respectively). A clause is satisfied if at least one of its literals is assigned to `true`. A SAT formula is a conjunction of clauses. We refer to a solution of a SAT formula as an assignment to its variables that satisfies all its clauses. Formulas with a solution are called *satisfiable*, and formulas without solutions are called *unsatisfiable*. Let $\vee$ and $\wedge$ refer to the logical OR and AND operators, respectively. For example, the formula $(x \vee \bar{y}) \wedge (\bar{x} \vee y)$ with two clauses is satisfiable. The solutions for this formula are the two assignments that assign both x and y to the same value.

SAT solvers, programs that solve SAT formulas, have become extremely powerful over the past two decades. Progress has been by leaps and bounds, starting with the pioneering work by Davis and Putnam [10] until the early 1990s when solvers could handle formulas with thousands of clauses. Today's solvers can handle formulas with millions of clauses. This performance boost resulted in the *SAT revolution* [4]: Encode problems arising from many interesting applications as SAT formulas, solve these formulas, and decode the solutions to obtain answers for the original problems. This is in a sense just using the *NP-completeness* of SAT [7, 13, 21]: Every problem with a notion of "solution"—where these solutions are relatively short and where an alleged solution can be verified (or rejected) quickly—can be reduced to SAT efficiently. For many years, NP-completeness was used only as a sign of "you cannot solve it!" but the SAT revolution has put this problem back on its feet.

Encoding

$(x_1 \vee x_2 \vee x_3) \wedge (\bar{x}_1 \vee \bar{x}_2 \vee \bar{x}_3) \wedge (x_1 \vee x_3 \vee x_4) \wedge (\bar{x}_1 \vee \bar{x}_3 \vee \bar{x}_4) \wedge$
$(x_1 \vee x_4 \vee x_5) \wedge (\bar{x}_1 \vee \bar{x}_4 \vee \bar{x}_5) \wedge (x_2 \vee x_3 \vee x_5) \wedge (\bar{x}_2 \vee \bar{x}_3 \vee \bar{x}_5) \wedge$
$(x_1 \vee x_5 \vee x_6) \wedge (\bar{x}_1 \vee \bar{x}_5 \vee \bar{x}_6) \wedge (x_2 \vee x_4 \vee x_6) \wedge (\bar{x}_2 \vee \bar{x}_4 \vee \bar{x}_6) \wedge$
$(x_1 \vee x_6 \vee x_7) \wedge (\bar{x}_1 \vee \bar{x}_6 \vee \bar{x}_7) \wedge (x_2 \vee x_5 \vee x_7) \wedge (\bar{x}_2 \vee \bar{x}_5 \vee \bar{x}_7) \wedge$
$(x_3 \vee x_4 \vee x_7) \wedge (\bar{x}_3 \vee \bar{x}_4 \vee \bar{x}_7) \wedge (x_1 \vee x_7 \vee x_8) \wedge (\bar{x}_1 \vee \bar{x}_7 \vee \bar{x}_8) \wedge$
$(x_2 \vee x_6 \vee x_8) \wedge (\bar{x}_2 \vee \bar{x}_6 \vee \bar{x}_8) \wedge (x_3 \vee x_5 \vee x_8) \wedge (\bar{x}_3 \vee \bar{x}_5 \vee \bar{x}_8) \wedge$
$(x_1 \vee x_8 \vee x_9) \wedge (\bar{x}_1 \vee \bar{x}_8 \vee \bar{x}_9) \wedge (x_2 \vee x_7 \vee x_9) \wedge (\bar{x}_2 \vee \bar{x}_7 \vee \bar{x}_9) \wedge$
$(x_3 \vee x_6 \vee x_9) \wedge (\bar{x}_3 \vee \bar{x}_6 \vee \bar{x}_9) \wedge (x_4 \vee x_5 \vee x_9) \wedge (\bar{x}_4 \vee \bar{x}_5 \vee \bar{x}_9)$

Case split as binary tree

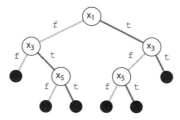

FIGURE 1. Encoding and case split of Boolean Schur triples problem. See also color image.

For many applications, including hardware and software verification [8, 20], SAT solving has become a disruptive technology that allows problems to be solved faster than by other known means.

The main paradigms of SAT solving are the incomplete *local search* [22], which can only find satisfying assignments, and the two complete paradigms (which can also determine unsatisfiability), *look-ahead* [19] and *conflict-driven clause learning* (CDCL) [30]. Local search tries to find a solution via local modifications to total assignments (using all variables). Look-ahead recursively splits the problem as cleverly as possible into subproblems, via looking ahead. CDCL tries to assign variables to find a satisfying assignment in a straightforward way, and if that fails (the normal case), then the failure is transformed into a clause, which is added to the formula. Here, we first explain CDCL, which is mainly responsible for the SAT revolution. Afterward, we describe how look-ahead can enhance CDCL on hard problems.

CDCL SAT solving algorithms cycle through three phases: *simplify, decide*, and *learn*. Solvers maintain an assignment (initially empty), and each phase updates that assignment. During *simplify*, the assignment is extended by detecting new inferences. Afterward, *decide* heuristically picks an unassigned variable and assigns it to `true` or `false`. After iterating these two phases, the current assignment either satisfies the formula, which terminates the search, or falsifies a clause. In the latter case, *learn* this conflict, as a clause, and modify the assignment to resolve the conflict. If the empty clause $\perp$ is learned, the solver detects unsatisfiability; otherwise, simplify-decide is performed again, etc. Look-ahead differs from CDCL by using stronger means for *simplify* and *decide*, but weaker means for *learn*.

The most basic inference mechanism in SAT solvers works as follows: A clause is *unit* under an assignment that falsifies all but one of its literals, while leaving the remaining literal unassigned. The only possibility to satisfy a unit clause (under that assignment) is to assign the remaining literal to `true`. A key SAT solving technique is *unit clause propagation* (UCP): Given an assignment and a formula, while the formula has unit clauses, extend the assignment by satisfying the remaining literals in the unit clauses. UCP has two possible terminating states: Either all unit clauses have been satisfied, or there is a falsified clause because of two complementary unit clauses (x) and $(\bar{x})$. In the latter case, we say that UCP results in a *conflict*. Conflicts are analyzed to obtain new clauses.

These *conflict clauses* are added to the formula to prevent the solver from visiting that assignment in the future. Additionally, conflict analysis updates the heuristics to guide the solver toward a short refutation.

There are two types of decision heuristics for SAT solvers: *focus* and *global* heuristics. Focus heuristics, also known as conflict-driven heuristics (for CDCL solvers), aim at finding short refutations. These heuristics are cheap to compute and have been highly successful in solving large problems arising from industrial applications. In short, focus heuristics work as follows: Whenever a solver encounters a conflicting state, the importance of the variables that cause the conflict is increased. Simply making these variables more important than all the other variables results in state-of-the-art performance on most industrial problems [3].

If no short refutation exists (or is too hard to find), it is best to use global heuristics (for look-ahead solvers) to split the search space into two parts that are both easier to solve. Global heuristics are based on *look-aheads* [25]: For a given formula F, a look-ahead on literal x assigns x to `true`, applies UCP, and computes the set S of clauses in F that are shortened but not satisfied. The heuristic value of a look-ahead on x is based on a weighted sum of the clauses in S, where clause weights depend on the length of clauses.

Both focus and global heuristics can reduce the search space exponentially. For really hard problems, such as the Pythagorean triples problem, it is best to combine both types of heuristics. Focus heuristics are effective when there exists a short refutation of the formula. For hard problems, initially there are no short refutations. One therefore needs to partition such a problem using global heuristics until the short refutations manifest themselves. This is the main idea behind the cube-and-conquer SAT solving paradigm [18], which was crucial to solve the Pythagorean triples problem.

Consider again the Boolean Schur triples problem on the existence of a red/blue coloring of 1, . . ., 9 without a monochromatic solution of $a + b = c$. Figure 1 shows the SAT encoding, consisting of 32 clauses using the Boolean variables $x_1, \ldots, x_9$. If variable x_i is assigned to `true` (`false`), then number i is colored red (blue). For each of the 16 solutions of $a + b = c$, there are two clauses: one stating that at least one of a, b, or c must be colored red, one stating that at least one of them must be colored blue. A binary tree is shown right beside the clauses. Each internal node contains a splitting variable x_i. The left branches assign decision

Proof	Unit clause justification
$(x_1 \vee x_3)$	$(\bar{x}_1 \vee x_2 \vee \bar{x}_3), (\bar{x}_1 \vee \bar{x}_3 \vee x_4), (\bar{x}_2 \vee \bar{x}_4 \vee \bar{x}_6), (\bar{x}_1 \vee \bar{x}_6 \vee x_7), (\bar{x}_3 \vee \bar{x}_6 \vee x_9), (\bar{x}_2 \vee \bar{x}_7 \vee \bar{x}_9)$
$(x_1 \vee x_5)$	$(\bar{x}_1 \vee x_3), (\bar{x}_1 \vee x_4 \vee \bar{x}_5), (\bar{x}_1 \vee \bar{x}_5 \vee x_6), (\bar{x}_2 \vee \bar{x}_4 \vee \bar{x}_6), (\bar{x}_2 \vee \bar{x}_5 \vee x_7), (\bar{x}_3 \vee \bar{x}_4 \vee \bar{x}_7)$
(x_1)	$(\bar{x}_1 \vee x_3), (\bar{x}_1 \vee x_5), (\bar{x}_2 \vee \bar{x}_3 \vee \bar{x}_5), (\bar{x}_3 \vee \bar{x}_5 \vee \bar{x}_6), (\bar{x}_2 \vee x_6 \vee \bar{x}_8), (\bar{x}_1 \vee \bar{x}_8 \vee x_9), (\bar{x}_3 \vee \bar{x}_6 \vee \bar{x}_9)$
$d(x_1 \vee x_3)$	
$d(x_1 \vee x_5)$	
$(\bar{x}_3)$	$(x_1), (\bar{x}_1 \vee \bar{x}_2 \vee \bar{x}_3), (\bar{x}_1 \vee \bar{x}_3 \vee \bar{x}_4), (\bar{x}_2 \vee \bar{x}_4 \vee x_6), (\bar{x}_1 \vee \bar{x}_6 \vee \bar{x}_7), (\bar{x}_3 \vee \bar{x}_6 \vee \bar{x}_9), (\bar{x}_2 \vee \bar{x}_7 \vee \bar{x}_9)$
$(\bar{x}_5)$	$(x_1), (\bar{x}_3), (\bar{x}_1 \vee \bar{x}_4 \vee \bar{x}_5), (\bar{x}_1 \vee \bar{x}_5 \vee \bar{x}_6), (x_2 \vee \bar{x}_4 \vee \bar{x}_6), (\bar{x}_2 \vee \bar{x}_5 \vee \bar{x}_7), (\bar{x}_3 \vee \bar{x}_4 \vee \bar{x}_7)$
$\bot$	$(x_1), (\bar{x}_3), (\bar{x}_5), (x_2 \vee \bar{x}_3 \vee \bar{x}_5), (\bar{x}_3 \vee \bar{x}_5 \vee x_8), (\bar{x}_2 \vee \bar{x}_6 \vee \bar{x}_8), (\bar{x}_1 \vee \bar{x}_8 \vee \bar{x}_9), (\bar{x}_3 \vee \bar{x}_6 \vee \bar{x}_9)$

FIGURE 2. Proof and unit clause justification of the Boolean Schur triples problem. See also color image.

variables to `false` (blue edge), while the right branches assign decision variables to `true` (red edge). Each leaf node represents an assignment that would result in a conflict during UCP. For example, for the left-most leaf node, x_1 and x_3 are assigned to `false` (blue): thus x_2, x_4 have to be set to `true` (because $1+2=3$ and $1+3=4$), forcing x_6 to `false` ($2+4=6$), which forces x_7 and x_9 to `true` ($1+6=7$ and $3+6=9$), which yields the conflict $2+7=9$ with all three set to `true` (red). This node matches the first clause in the proof of Figure 2. The binary tree (a simple form of look-ahead solving) illustrates that heuristics can reduce the number of assignments to be evaluated from 512 to 6.

Because of the limited size of the example formula, relatively simple heuristics are sufficient to reduce the number of cases from 512 to 6. One such simple heuristic is Maximum Occurrences in clauses of Minimal Size (MOMS). Initially, all clauses are ternary and variable x_1 occurs most frequently. Therefore, x_1 is used as the first decision variable. After simplification, several variables occur most frequently in binary clauses (twice), but variable x_3 has the best tie break (occurrences in remaining ternary clauses). Therefore, variable x_3 is the best decision on the second level of the tree. Finally, variable x_5 is the most occurring variable in binary clauses on the third level.

A crucial aspect of solving the Boolean Pythagorean triples problem was the use of a dedicated look-ahead heuristic based on the recursive weight heuristic for random 3-SAT formulas. The three magic constants in this heuristic have been manually tweaked to achieve strong performance on the Boolean Pythagorean triples problem [17]. We estimate that the use of this optimized look-ahead heuristic reduced the number of cases by at least two orders of magnitude compared to alternative heuristics, such as focus heuristics or MOMS. Look-ahead heuristics were popular in the 1990s, but they have been mostly ignored since CDCL emerged. The usefulness of look-ahead heuristics to boost performance on hard problems may revive interest.

Proofs of Unsatisfiability

The unpredictable effectiveness of SAT solvers, together with their nontrivial implementations (needed for real-world efficiency), raise the question of whether their results can be trusted. If a problem has a solution, it is easy to verify that the given solution is correct: simply check whether the solution satisfies at least one literal in every clause. However, a claim that no solution exists is much harder to validate. Since SAT solvers use many complicated techniques that could result in implementation as well as conceptual errors, a method is required to verify unsatisfiability claims.

There are two approaches to deal with the trust issue of complicated software: prove its correctness or produce a certificate which can be validated with a simple program. Work in the first direction resulted in verified SAT solving [33]. However, this approach has two disadvantages: Only some state-of-the-art techniques are verified, and verification is performed only on "higher levels," and thus excludes the low-level implementation tricks that are crucial for fast performance. Both disadvantages slow down the verified solver substantially, making it useless in most practical settings.

The second approach has been more successful in the context of SAT solving. We refer to a certificate of an unsatisfiability claim as a *proof of unsatisfiability*. What kind of format would be useful for such proofs? The ideal proof format facilitates five properties: (1) proof production should be *easy* to ensure that it will be supported by many solvers; (2) proofs should be *compact* in order to have small overhead; (3) proof validation should be *simple*; otherwise, the trust issue persists; (4) proof validation should be *efficient* to make verification useful in practice; and (5) all techniques should be *expressible*; otherwise, solvers will be handicapped. There is a trade-off among these properties. For example, more details in a proof should allow a more efficient validation procedure. However, adding details makes proofs less compact and harder to produce.

Initially, proofs of unsatisfiability were based on resolution. Although useful in some settings, it is hard or even impossible to achieve the properties of easy production (1), compactness (2), and expressibility (5) for such proofs. The alternative is *clausal proofs* [14], for which it is now possible to achieve all five properties.

What is a clausal proof of unsatisfiability for a SAT problem? Basically, we start with the given list of clauses, and add or delete clauses, until finally we add the empty clause $\bot$, which marks unsatisfiability, since there is no literal in it to satisfy. The most basic restriction on adding clauses is that the addition is *solution-preserving*, that is, all solutions (at that point, taking all previous additions and deletions into account) also satisfy the added clause. This step guarantees correctness: If all additions are solution-preserving, and we are able to add $\bot$ (which has no solution), then the original SAT problem must be unsatisfiable. For example, consider the formula $F = (x \vee y) \wedge (x \vee \bar{y})$. Adding the clause (x) to F is solution-preserving: F has two solutions, and in both solutions x is assigned to `true`.

It is important to validate that clause addition steps are solution-preserving; otherwise, we do not have a *proof*, just some sort of claim. This verification should be cheap to perform, and the basic criterion is as follows. Suppose a formula F is given, and the clause C is claimed to be solution-preserving for F. Take the assignment that sets all literals in C to `false`. If UCP on F results in a conflict, then the clause is indeed solution-preserving since we checked that it is not possible to falsify C while satisfying F. This method realizes the first three ideal proof format properties: easy, compact, and simple. The solver can just output the learned clauses, without a justification, and validation happens by UCP.

SAT solvers not only learn lots of clauses, but also aggressively delete them to achieve fast UCP. Proofs should include this deletion information in order to realize efficient validation. Furthermore, proof checkers require dedicated UCP algorithms to make proof validation as fast as proof production [16]. Combining these techniques realizes the fourth ideal proof property (efficient validation).

A proof of our running example is shown in Figure 2. The proof consists of six clause addition steps and two clause deletion steps. The latter have a "d" prefix and do not require checking. The correctness of each clause addition step is checked using UCP, and shown using a unit clause justification: a sequence of clauses that become unit, ending with a falsified clause that marks the conflict. The unit clause justification is omitted from clausal proofs to ensure compactness, but the checker constructs a justification during validation.

Some SAT solving techniques may change (add or remove) solutions, which can significantly reduce solving time. In order to express such techniques—to have also the final ideal proof property (expressible)—support is required for proof steps that go beyond the above solution preservation. This support is realized by the concept of *solution-preserving modulo x* for some literal x. Let φ be an assignment. We denote by $\varphi \oplus x$ the assignment obtained by flipping the truth value for literal x in φ. In case x is unassigned in φ, then x is assigned to `true` in $\varphi \oplus x$. For a given formula F, addition of clause C is solution-preserving modulo x if for all solutions φ of F at least one of φ or $\varphi \oplus x$ satisfies *F and C*.

For example, consider the formula $F = (x \lor y) \land (x \lor \bar{y})$ again. The addition of clause $(\bar{x} \lor y)$ to F is solution-preserving modulo y. Recall that F has two solutions. The first solution φ_1, where x is `true` and y is `true`, also satisfies $(\bar{x} \lor y)$. The second solution φ_2, where x is `true` and y is `false`, falsifies $(\bar{x} \lor y)$, but $\varphi_2 \oplus y$ satisfies F and $(\bar{x} \lor y)$.

How to check that adding clause C is solution-preserving modulo x? We use the following efficient criterion: $x \in C$, and for all $D \in F$ with $\bar{x} \in D$, we have that setting all literals in C *as well as* all literals in $D \backslash \{\bar{x}\}$ to `false` yields a conflict via UCP. The proof format that encapsulates this inference in a single step is called the "DRAT" format [4], and it is supported by state-of-the-art solvers.

It is instructive to show that this criterion guarantees that adding C to F is solution-preserving modulo x. The critical clauses are the $D \in F$ with $\bar{x} \in D$, since here flipping of x might change a satisfied clause to a falsified clause. First observe that from the criterion follows that all $C \cup (D \backslash \{\bar{x}\})$ are solution-preserving with respect to F. Now assume that φ is a total satisfying assignment for F, which falsifies C (otherwise, φ satisfies F and C, and we are done). Thus, φ falsifies x, and $\varphi \oplus x$ satisfies C. Since all $C \cup (D \backslash \{\bar{x}\})$ are solution-preserving with respect to F, φ satisfies all $C \cup (D \backslash \{\bar{x}\})$. Hence, φ satisfies all $D \backslash \{\bar{x}\}$ (because φ falsifies C), and so does $\varphi \oplus x$ as well, and thus indeed $\varphi \oplus x$ satisfies all D. QED.

The DRAT format seems to be a good proof format for existing and future SAT solvers, as it has all the five properties of an ideal proof format. Moreover, DRAT proofs can be efficiently checked even in parallel, and they have been used to validate the results of the annual international SAT competitions since 2013. For the Boolean Schur triples problem with $n = 9$, there exists a DRAT proof consisting of only four clause additions: $(x_1 \vee x_4)$, (x_1), (x_4), $\perp$. Validating this proof involves more details, which can be obtained by using the DRAT proof checker DRAT-TRIM.[4]

Indeed, DRAT in a theoretical sense is equivalent to one of the most powerful systems studied in proof complexity, extended Frege with substitution, and thus it should offer "proofs as short as possible" [5]. The extension rule basically states that the clauses $(x \vee \bar{a} \vee \bar{b}) \wedge (\bar{x} \vee a) \wedge (\bar{x} \vee b)$ can be added if no literals x and $\bar{x}$ occur in the formula. In fact, each of the clauses are solution-preserving modulo x or $\bar{x}$ according to the above criterion.

Proof size nevertheless becomes an issue. Although DRAT proofs are "compact," the size of the DRAT proof of the Boolean Pythagorean triples problem is 200 TB. An obvious challenge of such a huge file is its storage. Also, dealing with such files increases the complexity of proof validation algorithms, which will need to support parallel checking. On the other hand, it is possible to trade complexity for space by adding details to the proof that facilitate fast checking. In order to make this feasible, the proof can be optimized using a nonverified trimmer, which also adds the checking details. This approach has been successfully applied to validate the 200 TB proof using a checker that was *formally verified* in CoQ [9].

Ramsey Theory and Complexity

A popularized summary of Ramsey theory is that "complete chaos is impossible" [28]. More concretely, Ramsey theory deals with patterns that occur in well-known sets such as the set of natural numbers or the set of graphs. For example, coloring the natural numbers with finitely many colors will result in a monochromatic Schur triple $a + b = c$.

Hundreds of papers have been published on determining the smallest size of sets such that a given pattern must start to occur [34]. The most famous pattern is related to Ramsey numbers $R(k)$: the smallest n such that all red/blue edge colorings of the complete graph with n vertices have a red or a blue clique of size k. Only the first four Ramsey numbers are known. Paul Erdős famously told a story about aliens who threatened to obliterate Earth unless humans provided them with the value of

$R(5)$—with a proof, we may add here. Putting all mankind behind this project would do the job in a year. Yet if aliens asked for $R(6)$, we should opt for the Hollywood resolution and obliterate them instead [15].

Many problems in Ramsey theory appear to be solved only using large case-splits (especially for the determination of Ramsey-type numbers), and thus using SAT is a natural option. Also SAT formulations of these problems are easy and natural. In order to determine the smallest subset in which a pattern starts to occur using SAT, two formulas need to be solved. First, it has to be shown that for any smaller subset there exists a counterexample. This step is typically easy because the formula is satisfiable. The second formula, encoding the existence of the pattern, is much harder to solve as now unsatisfiability must be shown.

The first major success of SAT solving in Ramsey theory was determining the sixth Boolean van der Waerden number [24]: $\mathrm{vdW}(6) = 1{,}132$. The number $\mathrm{vdW}(k)$ expresses the smallest n such that any red/blue coloring of the numbers 1 to n results in a monochromatic arithmetic progression of length k. The computation used multiple clusters as well as dedicated SAT-solving hardware (field-programmable gate array, or FPGA, solvers) for several months. Unfortunately, no proof was produced during the computation, making it impossible to verify the result. This situation raises several trust issues because errors could have been made on several levels. For example, was the splitting correct and thus has the whole search space been explored? Also, FPGA solvers have been tested much less thoroughly compared to state-of-the-art solvers.

The first important problem with a verified clausal proof is the Erdős discrepancy problem (EDP), which states that "complete uniformity is impossible." The problem conjectures that any infinite sequence $s_1, s_2, \ldots$ with $s_i = \pm 1$ contains for any positive integer C a subsequence $s_d, s_{2d}, s_{3d}, \ldots, s_{kd}$, for some positive integers k and d, such that $|\sum_{i=1}^{k} S_{id}| \geq C$. Using colors, the conjecture says that for every $C \geq 1$ and every red/blue coloring of 1, 2, . . ., there is a finite initial segment of some progression d, $2d$, $3d$, . . . for some $d \geq 1$, such that the discrepancy between the number of color occurrences is at least C (one color occurs at least C times more than the other). The conjecture has been a long-standing open problem even for $C = 2$. The case $C = 2$ was eventually solved using SAT by providing the exact bound [23] and also applying cube-and-conquer. The encoding of this problem is more involved than the simple encoding of Ramsey problems (which are just hypergraph coloring problems), and thus, though a clausal proof has been provided, correctness is more of an issue than in cases of Ramsey theory. Computationally, EDP is much easier [23], and a much smaller proof exists (about a gigabyte) than in our case. Finally, a general mathematical existence proof has been provided [37]. This mathematical proof was called "much more satisfying" than the computational approach [27]. However, there is for example the possibility that the Pythagorean tuples conjecture (see below) is not provable with current methods. Furthermore, the SAT approach is actually a rather "satisfying approach" when taking into account its deep connections to formal methods.

The *Pythagorean tuples conjecture* states that $Ptn(k; m)$—with k the length of the tuple and m the number of colors—exists for all $k \geq 3$ and $m \geq 2$. That is, for every partitioning of $\{1, \ldots, Ptn(k; m)\}$ into m parts, some part contains a Pythagorean tuple of size k. We have shown that $Ptn(3; 2) = 7,825$. The value of $Ptn(3; 2)$ was conjectured [32] not to exist after determining the numbers $Ptn(k; 2)$ for $4 \leq k \leq 31$. We have meanwhile computed the only known Pythagorean tuples numbers for three colors: $Ptn(5; 3) = 191$, $Ptn(6; 3) = 121$, and $Ptn(7; 3) = 102$. We also established $Ptn(3; 3) > 10^7$, and this lower bound (via local-search algorithms) seems still far away from the exact bound. So it is imaginable that a mathematical existence proof cannot be found, and finiteness of $Ptn(3; 3)$ might never be established. It is furthermore conceivable that the Pythagorean tuples conjecture is true but the best proofs are SAT-like. Thus, formal proofs in systems like Zermelo-Fraenkel set theory would only *exist* for concrete k and m, while there would not exist a single proof for all k and m. No mathematical existence proofs have yet been established for any $Ptn(k; m)$ (see "alien truth statements" for further discussions).

Before coming to the industrial applications of SAT, we remark that the Ramsey numbers [35] $R(k)$ are very different from the Boolean Pythagorean triples problem: namely, the latter is "random-like" and thus has no symmetries (besides the trivial color symmetries). Currently, SAT solving is more successful in the absence of strong symmetries, and Ramsey numbers currently have too much structure for an automated attack. More sophisticated symmetry-breaking techniques are required to improve the performance.

Brute Force Formal Methods

SAT solvers are a key technology in formal methods for applications, such as bounded model checking [6] and equivalence checking. In bounded model checking, given a transition system and an invariant such as a safety property, the SAT solver determines for some appropriate finitization, whether there exists a sequence of transitions that violates the safety property. Equivalence checking is used to determine the equivalence of a specification and an implementation or two different implementations. The SAT solver is asked to find an input such that some output differs. Notice that the existence of a solution means that the safety property is violated or that there exists a counterexample for equivalence.

All problems discussed so far could be expressed as a propositional formula. For many interesting problems, however, this is not the case and they require a richer logic for its representation. That does not mean that SAT technology cannot be used to solve these problems. On the contrary, more and more problems that require a richer logic are being solved efficiently using SAT.

The key idea is to abstract away those parts of a given problem that cannot be expressed as propositional logic. A solution of the abstracted problem may not be a solution of the given problem, while a refutation of the abstracted problem is also a refutation of the given problem. In case a solution of the abstracted problem is obtained, which is not a solution for the given problem, then the abstraction is refined by adding a clause that prevents the SAT solver from finding that solution (and potentially similar solutions) again. This sequence is repeated until either a refutation or a solution for the given problem is found. Incremental SAT solving [12] facilitates an efficient implementation of this approach.

This approach has been very successful in automated theorem proving (ATP). The long-time champion in the annual ATP competitions is VAMPIRE [38], which has been tightly integrated with a SAT solver. Other strong ATP solvers, including IPROVER and LEO, incorporate SAT solvers as well. The major interactive theorem provers, such as ACL2, COQ, and ISABELLE, support the usage of SAT solvers to deal with subproblems that can be expressed in propositional logic. In this setting, each SAT solver is treated as a black box, and the emitted proofs are validated in the theorem provers. Another successful extension of SAT in this direction is *satisfiability modulo theories* (SMTs) [11].

SMTs use multiple theories (such as linear arithmetic, uninterpreted functions, and bit vectors) and replace constraints in a theory by propositional variables. SMT solvers, such as Z3, BOOLECTOR, CVC4, and YICES have been highly successful.

Alien Truths

The core argument against solving a problem by brute force is that it does not contribute to understanding the problem. In that view, the proof is meaningless and hard to generalize, and a human mathematical proof is preferred. Furthermore, without understanding, errors seem more likely, although validation can be done by highly trusted systems.

The proponents of "elegant" proofs appear to consider problems with only very long proofs as not interesting or not relevant. But even unprovable statements, like the famous continuum hypothesis, have an important place in mathematics. If we do not study the limits of our current knowledge, we will stay ignorant forever, always restricted to a "safe space," neglecting problems we *assume* to be too hard. Furthermore, what is a limit of one discipline is a core subject of another discipline. Computational complexity and Ramsey theory have close relations. *Understanding* the hardness of problems from Ramsey instances could lead to major breakthroughs [29]. For example, why is proving the Ramsey property for $a + b = c$ rather easy, whereas $a^2 + b^2 = c^2$ appears to be a very hard problem? In general, even small propositional problems might have only very large proofs. If we were to ignore this area, then we would allow random holes in our knowledge. The question "why there are *no* short proofs" and "what makes a problem hard" are deep and fascinating questions, and we consider them some of the most important problems of our times.

To better discuss the untold stories of computer science, complexity theory, and SAT, let's call *alien* a provable and rather short mathematical statement with only a very long proof. Artificial alien statements can be constructed using Gödel's methods. Whether a natural truth statement can be shown to be alien, such as the Pythagorean triples problem, is of highest relevance. Even if a short proof for the Pythagorean triples problem may be constructed, that is unlikely to be the case for the exact bound result. Now there is actually a whole spectrum of possibilities between human truths and alien truths. Classical mathematical statements for which a paper proof exists, such as Schur's theorem [36], we consider as *human* truth statements. Hence, the vast body of mathematical works belongs to this category. Furthermore, we consider mathematical statements that have been proven mostly manually, but with some computer help, *weakly human*. More specifically, such statements have a large case-split, which could potentially be understood by humans, but which have only been checked mechanically. An example of such a statement is the four-color theorem [39]. The proof by Appel and Haken [2] considers 663 cases in its improved version. The case-split is fully understood and humanly constructed. A theorem prover only checks the cases. Coming to larger cases, we refer to a *weakly alien* truth statement as a giant humanly generated case-split that can be validated using plain brute force methods. For example, it has been shown that the minimum number of givens is 17 in Sudoku by enumerating all possible cases with 16 givens and refuting them all [31] (5,472,730,538 cases after symmetry breaking). Although impossible to evaluate by humans, it could be directly done mechanically. This result is expected to be weakly alien, as it is unlikely that there exists a small enough case-split that is checkable by humans.

We arrive at a better understanding of "alien," namely a truth statement is *alien* if humanly understandable case-splits are way too big for any plain brute force method, but there exists a giant case-split that mysteriously avoids an enormous exponential effort. Examples of truth statements that are expected to be alien are that vdW(6) = 1,132 (see Kouril and Paul [24]) and that the exact bound of EDP with $C = 2$ is 1,161 (see Konev and Lisitsa [23]). A plain brute force approach to those problems would require the evaluation of 2^{1132} and 2^{1161} cases, respectively. Brute reasoning using SAT solvers significantly reduced the size of the case-splits and allowed determining their truth. We think it is relevant to make a further distinction: the above two alien truth statements express the exact bound, but for both cases there is a mathematical existence proof that the pattern cannot be avoided indefinitely. Now also high-level statements, such as any red/blue coloring of the natural numbers yields a monochromatic Pythagorean triple, could be alien, when the bound result, Ptn(3; 2) = 7,825, is the only proof. We call such statements indeed *strongly alien*. If a mathematical existence proof is found for the statement here, then only the bound statement remains, which is simply *alien*. This happened for the Erdős discrepancy problem: The bound was computed using SAT, and later a mathematical existence proof was given.

Finally, for some truth statements, we may never be able to produce a proof. A possible example problem of this type is the statement that every 3-coloring of the natural numbers yields a monochromatic Pythagorean triple. As already discussed, experiments show that $Ptn(3; 3) > 10^7$, where lower bounds are relatively easy to compute. Proofs of upper bound results are much harder to obtain: for example, $Ptn(3; 2) > 7,824$ can be computed in one CPU-minute with a local search, whereas computing $Ptn(3; 2) \leq 7,825$ required more than 40,000 CPU-hours. We call decidable truth statements *extra-alien* if a proof can never be computed.

The concept of alien truth statements deals with the *size* of proofs, but it touches naturally on *unprovability* (in current systems like Zermelo-Fraenkel set theory). It is conceivable that $Ptn(3; 3)$ does not exist; that is, the natural numbers are 3-colorable without a monochromatic Pythagorean triple. However, this statement may not be provable, since the coloring is too complex. On the other hand, it is conceivable that all $Ptn(3; m)$ with $m \geq 3$ exist (note that a SAT solver can prove them in principle), but these statements are all alien or extra-alien. Since these proofs grow with m, the general statement that all $Ptn(3; m)$ with $m \geq 3$ exist, is then unprovable *in principle*.

Conclusion

Recent successes in brute reasoning, such as solving the Erdős discrepency problem and the Pythagorean triples problem, show the potential of this approach to deal with long-standing open mathematical problems. Moreover, proofs for these problems can be produced and verified completely automatically. These proofs may be big, but we argued that compact elegant proofs may not exist for some of these problems, in particular (but not only) for the exact bound results. The size of these proofs does not influence the level of correctness, and these proofs may reveal interesting information about the problem.

In contrast to popular belief, mechanically produced huge proofs can actually help in understanding the given problem. We can try to understand their structure, making them thus smaller. Hardly any

research has been done yet in this direction apart from removing re-
dundancy in a given proof. Possibilities are changing the heuristics
of a solver or introducing new definitions of frequently occurring
patterns in the proof. Indeed, simply validating a clausal proof does
not only produce a yes/no answer as to whether the proof is correct, but also provides an *unsatisfiable core* consisting of all original
clauses that were used to validate the proof—revealing important
parts of the problem. The size of the core depends on the type of
problem. Problems in Ramsey theory typically have quite a large core
and therefore provide limited insight. Many bounded model checking
problems, however, have small unsatisfiable cores, thereby showing
that large parts of the hardware design were not required to determine the safety property.

To conclude, it is definitely possible to gain insights by using SAT.
However that "insight" might need to be reinterpreted here and might
work on a higher level of abstraction. Every paradigm change means
asking different questions. Gödel's incompleteness theorem *solved* partially the question of the consistency of mathematics by showing that
the answer provably cannot be delivered in the naive way. Now the task
is to live up to big complexities and to embrace the new possibilities.
Proofs must become objects for investigations, and understanding will
be raised to the next level, how to find and handle them.

So, when the day finally comes and the aliens arrive and ask us
about Ptn(3; 3), we will tell them: "You know what? Finding the
answer yourself gives you a much deeper understanding than just telling you the answer—here you have the SAT solving methodology;
that's the real stuff!" And then humans and aliens will live happily
ever after.

Wir müssen wissen. Wir werden wissen.
(We must know. We will know.)

—David Hilbert, 1930

Notes

1. https://rjlipton.wordpress.com/the-gdel-letter/.
2. http://fmv.jku.at/lingeling/.
3. http://www.satcompetition.org/.
4. The tool is available at https://github.com/marijnheule/drat-trim.

References

1. Ahmed, T., Kullmann, O., Snevily, H. On the van der Waerden numbers w(2; 3, *t*). *Disc. Appl. Math. 174* (2014), 27–51.

2. Appel, K., Haken, W. Every planar map is four colorable. Part I: Discharging. *Illinois J. Math. 21*, 3 (1977), 429–90.

3. Biere, A., Fröhlich, A. Evaluating CDCL variable scoring schemes. In *SAT* (Springer, 2015), 405–22.

4. Biere, A., Heule, M.J.H., van Maaren, H., Walsh, T., eds. *Handbook of Satisfiability*, volume 185 of *FAIA*. IOS Press, Amsterdam, Netherlands, Feb. 2009.

5. Buss, S. Propositional proofs in Frege and Extended Frege systems (abstract). In *Computer Science—Theory and Applications* (Springer, 2015), 1–6.

6. Clarke, E. M., Biere, A., Raimi, R., Zhu, Y. Bounded model checking using satisfiability solving. *Formal Methods in System Design 19*, 1 (2001), 7–34.

7. Cook, S. A. The complexity of theorem-proving procedures. In *STOC* (1971), 151–58.

8. Copty, F., Fix, L., Fraer, R., Giunchiglia, E., Kamhi, G., Tacchella, A., Vardi, M. Y. Benefits of bounded model checking at an industrial setting. In *CAV* (Springer, 2001), 436–53.

9. Cruz-Filipe, L., Marques-Silva, J. P., Schneider-Kamp, P. Efficient certified resolution proof checking, 2016. https://arxiv.org/abs/1610.06984.

10. Davis, Martin, Putnam, Hilary. A computing procedure for quantification theory. *J. ACM 7*, 3 (1960), 201–15.

11. de Moura, L., Bjørner, N. Satisfiability modulo theories: Introduction and applications. *Communications of the ACM 54*, 9 (2011), 69–77.

12. Eén, N., Sörensson, N. Temporal induction by incremental SAT solving. *Electr. Notes Theor. Comput. Sci. 89*, 4 (2003), 543–60.

13. Garey, M. R., Johnson, D. S. *Computers and Intractability; A Guide to the Theory of NP-Completeness*. W. H. Freeman and Company, 1979.

14. Goldberg, E. I., Novikov, Y. Verification of proofs of unsatisfiability for CNF formulas. In *DATE* (IEEE, 2003), 10886–91.

15. Graham, R. L., Spencer, J. H. Ramsey theory. *Scientific American 263*, 1 (July 1990), 112–17.

16. Heule, M.J.H., Hunt, W. A., Jr., Wetzler, N. Trimming while checking clausal proofs. In *FMCAD* (IEEE, 2013), 181–88.

17. Heule, M.J.H., Kullmann, O., Marek, V. W. Solving and verifying the Boolean Pythagorean triples problem via cube-and-conquer. In *SAT* (Springer, 2016), 228–45.

18. Heule, M.J.H., Kullmann, O., Wieringa, S., Biere, A. Cube and conquer: Guiding CDCL SAT solvers by lookaheads. In *HVC* (Springer, 2011), 50–65.

19. Heule, M.J.H., van Maaren, H. Look-ahead based SAT solvers. In Biere et al. [4], Chapter 5, (2009), 155–84.

20. Ivančić, F., Yang, Z., Ganai, M. K., Gupta, A., Ashar, P. Efficient SAT-based bounded model checking for software verification. *Theoretical Computer Science 404*, 3 (2008), 256–74.

21. Karp, R. M. Reducibility among combinatorial problems. In *Complexity of Computer Computations* (Plenum Press, 1972), 85–103.

22. Kautz, H. A., Sabharwal, A., Selman, B. Incomplete algorithms. In Biere et al. [4], Chapter 6, (2009), 185–203.

23. Konev, B., Lisitsa, A. Computer-aided proof of Erdős discrepancy properties. *Artificial Intelligence 224*, C (July 2015), 103–18.

24. Kouril, M., Paul, J. L. The van der Waerden number $W(2, 6)$ is 1132. *Experimental Mathematics 17*, 1 (2008), 53–61.

25. Kullmann, O. Fundaments of branching heuristics. In Biere et al. [4], Chapter 7, (2009), 205–44.

26. Lam, C.W.H. The search for a finite projective plane of order 10. *The American Mathematical Monthly 98*, 4 (April 1991), 305–18.

27. Lamb, E. Maths proof smashes size record: Supercomputer produces a 200-terabyte proof—but is it really mathematics? *Nature 534* (June 2016), 17–18.

28. Landman, B. M., Robertson, A. *Ramsey Theory on the Integers*, volume 24 of *Student mathematical library*. American Mathematical Society, Providence, RI, 2003.

29. Lauria, M., Pudlák, P., Rödl, V., Thapen, N. The complexity of proving that a graph is Ramsey. In *ICALP* (Springer, 2013), 684–95.

30. Marques-Silva, J. P., Lynce, I., Malik, S. Conflict-driven clause learning SAT solvers. In Biere et al. [4], Chapter 4, (2009), 131–53.

31. McGuire, G., Tugemann, B., Civario, G. There is no 16-clue Sudoku: Solving the Sudoku minimum number of clues problem via hitting set enumeration. *Experimental Mathematics 23*, 2 (2014), 190–217.

32. Myers, K. J. Computational advances in Rado numbers. Ph.D. thesis, Rutgers University, New Brunswick, NJ, 2015.

33. Oe, D., Stump, A., Oliver, C., Clancy, K. versat: A verified modern SAT solver. In *VMCAI* (Springer, 2012) 363–78.

34. Radziszowski, S. P. Small Ramsey numbers. *The Electronic Journal of Combinatorics* (January 2014), Dynamic Surveys DS1, Revision 14.

35. Ramsey, F. P. On a problem of formal logic. *Proceedings of the London Mathematical Society 30* (1930), 264–86.

36. Schur, I. Über die Kongruenz $x^m + y^m = z^m \pmod{p}$. *Jahresbericht der Deutschen Mathematiker-Vereinigung 25* (1917), 114–16.

37. Tao, T. The Erdős discrepancy problem. *Discrete Analysis 1* (February 2016), 29.

38. Voronkov, A. AVATAR: The architecture for first-order theorem provers. In *CAV* (Springer, 2014), 696–710.

39. Wilson, R. *Four Colors Suffice: How the Map Problem Was Solved*. Princeton University Press, Princeton, NJ, revised edition, 2013.

Computational Thinking in Science

Peter J. Denning

A quiet but profound revolution has been taking place throughout science. The computing revolution has transformed science by enabling all sorts of new discoveries through information technology.

Throughout most of the history of science and technology, there have been two types of characters. One is the experimenter, who gathers data to reveal when a hypothesis works and when it does not. The other is the theoretician, who designs mathematical models to explain what is already known and uses the models to make predictions about what is not known. The two types interact with one another because hypotheses may come from models, and what is known comes from previous models and data. The experimenter and the theoretician were active in the sciences well before computers came on the scene.

When governments began to commission projects to build electronic computers in the 1940s, scientists began discussing how they would use these machines. Nearly everybody had something to gain. Experimenters looked to computers for data analysis—sifting through large data sets for statistical patterns. Theoreticians looked to them for calculating the equations of mathematical models. Many such models were formulated as differential equations, which considered changes in functions over infinitesimal intervals. Consider, for example, the generic function f over time (abbreviated $f(t)$). Suppose that the differences in $f(t)$ over time give another equation, abbreviated $g(t)$. We write this relation as $df(t)/dt = g(t)$. You could then calculate the approximate values of $f(t)$ in a series of small changes in time steps, abbreviated Δt, with the difference equation $f(t+\Delta t) = f(t) + \Delta t g(t)$. This calculation could easily be extended to multiple space dimensions with difference equations that combine values on neighboring nodes of a grid. In his collected works, John von Neumann, the polymath who helped design

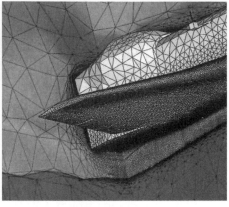

FIGURE 1. Aeronautics engineers use simulations from computational fluid dynamics to model airflows around proposed aircraft. They have become so good at this that they can test new aircraft designs without wind tunnels or test flights. The first step is to build a three-dimensional mesh in the space surrounding the aircraft (in this case, for the Space Shuttle). The spacing of the grid points is smaller near the fuselage, where the changes in air movement are greatest. Then the differential equations of airflow are converted to difference equations on the mesh. A supercomputer grinds out the profiles of the flow field and the forces on each part of the aircraft over time. The numerical results are converted to colored images (left) that reveal where the stresses on the aircraft are greatest. (Image at left courtesy of NASA; image at right courtesy of Peter A. Gnoffo and Jeffery A. White/NASA.). See also color image.

the first stored program computers, described algorithms for solving systems of differential equations on discrete grids.

Using the computer to accelerate the traditional work of experimenters and theoreticians was a revolution of its own. But something more happened. Scientists who used computers found themselves routinely designing new ways to advance science. Simulation is a prime example. By simulating airflows around a wing with a type of equation (called Navier-Stokes) that is broken out over a grid surrounding a simulated aircraft, aeronautical engineers largely eliminated the need for wind tunnels and test flights (Figure 1). Astronomers similarly simulated the collisions of galaxies, and chemists simulated the deterioration of space probe heat shields on entering an atmosphere. Simulation allowed scientists to reach where theory and experiment could not. It became a new way of doing science. Scientists became computational designers as well as experimenters and theoreticians.

Another important example of how computers have changed how science is done has been the new paradigm of treating a physical process as an information process, which allows more to be learned about the physical process by studying the information process. Biologists have made significant advances with this technique, notably with sequencing and editing genes. Data analysts also have found that deep learning models enable them to make surprisingly accurate predictions of processes in many fields. For the quantities predicted, the real process behaves as an information process.

The two approaches are often combined, such as when the information process provides a simulation for the physical process it models.

The Origins of a Term

The term *computational science*, and its associated term *computational thinking*, came into wide use during the 1980s. In 1982, theoretical physicist Kenneth Wilson received a Nobel Prize in physics for developing computational models that produced startling new discoveries about phase changes in materials. He designed computational methods to evaluate the equations of renormalization groups and used them to observe how a material changes phase, such as the direction of the magnetic force in a *ferrimagnet* (in which adjacent ions have opposite but unequal charges). He launched a campaign to win recognition and respect for computational science. He argued that all scientific disciplines had very tough problems—"grand challenges"—that would yield to massive computation. He and other visionaries used the term *computational science* for the emerging branches of science that used computation as their primary method. They saw computation as a new paradigm of science, complementing the traditional paradigms of theory and experiment. Some of them used the term *computational thinking* for the thought processes in doing computational science—designing, testing, and using computational models. They launched a political movement to secure funding for computational science research, culminating in the High-Performance Communication and Computing (HPCC) Act passed in 1991 by the U.S. Congress.

It is interesting that computational science and computational think-ing in science emerged from within the scientific fields—they were not imported from computer science. Indeed, computer scientists were slow to join the movement. From the beginnings of computer science in the 1940s, there was a small but important branch of the field that specialized in numerical methods and mathematical software. These computer scientists have the greatest affinity for computational science and were the first to embrace it.

Computation has proved so productive for the advancement of science and engineering that virtually every field of science and en-gineering has developed a computational branch. In many fields, the computational branch has grown to constitute the majority of the field. For example, in 2001, David Baltimore, Nobel laureate in biology, said that biology is an information science. Most recent advances in biology have involved DNA modeling, sequencing, and editing. We can expect this trend to continue, with computation invading deeper into every field, including social sciences and the humanities (Figure 2). Many people will learn to be computational designers and thinkers.

What Is Computational Thinking?

Computational thinking is generally defined as the mental skills that facilitate the design of automated processes. Although this term traces back to the beginnings of computer science in the 1950s, it became popular after 2006 when educators undertook the task of helping all children become productive users of computation as part of STEM (sci-ence, technology, engineering, and mathematics) education. If we can learn what constitutes computational thinking as a mental skill, we may be able to draw more young people to science and accelerate our own abilities to advance science. The interest from educators is forcing us to be precise in determining just what computational thinking is.

Most published definitions to date can be paraphrased as follows: "Computational thinking is the thought processes involved in formulating problems so that their solutions are represented as computational steps and algorithms that can be effectively carried out by an information-processing agent." This definition, however, is fraught with problematic ideas. Consider the word "formulating." People regularly formulate re-quests to have machines do things for them without having to understand

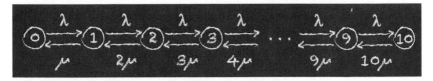

FIGURE 2. As an example of a problem aided by computational thinking, consider a telephone switching office. To determine its capacity, telephone engineers pick a target probability of overflow—for example, 0.001. They ask, What is the maximum number N of simultaneous phone calls so that the chances that a new caller cannot get a dial tone is less than 0.001? A random walk computational model yields an answer. The model has states $n = 0, 1, 2, \ldots, N$, representing the number of calls in progress up to a maximum of N; here $N = 10$. Requests to initiate new calls are occurring at rate λ. Individual callers hang up at rate μ. Each new-call arrival increases the state by 1, and each hangup decreases it by 1. The movement through the possible states is represented by the state diagram above. Telephone engineers define $p(n)$ as the fraction of time the system is in state n and can prove a difference equation $p(n) = (\lambda/n\mu)p(n-1)$. They calculate all the $p(n)$ by guessing $p(0)$ and then normalizing so that the sum of $p(n)$ is 1. Then they find the largest N so that $p(N)$ is below the target threshold. For example, if they find $p(N) = 0.001$ when $N = 10$, they predict that a new caller has a chance 0.001 of not getting a dial tone when the exchange capacity is 10 calls. See also color image.

how the computation works or how it is designed. The term "information agent" is also problematic—it quickly opens the door to the false belief that step-by-step procedures followed by human beings are necessarily algorithms. Many people follow "step-by-step" procedures that cannot be reduced to an algorithm and automated by a machine. These fuzzy definitions have made it difficult for educators to know what they are supposed to teach and how to assess whether students have learned it.

And what "thought processes" are involved? The published definitions say that they include making digital representations, sequencing, choosing alternatives, iterating loops, running parallel tasks, abstracting, decomposing, testing, debugging, and reusing. But this is hardly a complete description. To be a useful contributor, a programmer also needs to understand enough of a scientific field to be able to express problems and solution methods appropriate for the field. For example,

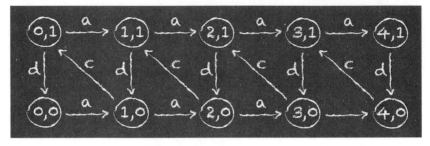

FIGURE 3. Computational design helps a doctor build an electronic controller for her office, which consists of a waiting room and a treatment room that holds four people. Patients enter the waiting room and sit down. As soon as the doctor is free, she calls the next patient into the treatment room. When done, the patient departs by a separate door. The doctor wants an indicator lamp to glow in the treatment room when patients are waiting, and another to glow in the waiting room when she is busy treating someone. The engineer designing the controller uses a computational model with states (n, t) where $n = 0, 1, 2, 3, 4$ is the number of patients in the waiting room and $t = 0, 1$ is the number of patients in the treatment room. The indicator lamp in the treatment room glows whenever $n > 0$, and the lamp in the waiting room glows whenever $t > 0$. The controller implements the state diagram above. State transitions occur at three events: patient arrival (a), patient departure (d), and patient call by the doctor (c). These events are signaled by sensors in the three doors. See also color image.

I once witnessed that a team of computational fluid dynamics scientists invited Ph.D. computer scientists to work with them, only to discover that the computer scientists did not understand enough fluid dynamics to be useful. They were not able to think in terms of computational fluid dynamics. The other team members wound up treating the computer scientists like programmers rather than peers, much to their chagrin. It seems that the thought processes of computational thinking should include those of skilled practitioners of the field where the computation will be used.

All these difficulties suggest that the word "thinking" is not what we are really interested in—we want the ability to design computations. Design includes the dimensions of listening to the community of users, testing prototypes to see how users react, and making technology offers that take care of user concerns (Figure 3). Therefore, *computational design* is a more accurate term. It is clearly a skill set, not a body of mental knowledge about programming.

What Is a Computational Model?

An essential aspect of computational design (or thinking) is a machine that will carry out the automated steps. But most computational designers do not directly consider the hardware of the machine itself; instead they work with a *computational model*, which is an abstract machine—basically a layer of software on top of the hardware that translates a program into instructions for the hardware. Designers are not concerned with mapping the model to the real machine because that's a simulation job that software engineers handle.

In computing science, the model most talked about is the Turing machine, which was invented in 1936 by computing pioneer Alan Turing. His model consists of an infinite tape and a finite state control unit that moves one square at a time back and forth on the tape, reading and changing symbols. Turing machines are the most general model for computation—anything that people reasonably think can be computed, can be computed by a Turing machine (Figure 4).

But Turing machines are too primitive to easily represent everyday computation. With each new programming language, computer scientists defined an associated abstract machine that represented the entity programmed by the language. Software called a *compiler* then translated the language operations on the abstract machine into machine code on the real hardware.

The models of the Turing machine and of programming languages are all general purpose—they deal with anything that can be computed. But we often work with much less powerful models that are still incredibly useful. One of the most common is the *finite state machine*, which consists of a logic circuit, a set of flip-flop switch circuits to record the current state, and a clock whose ticks trigger state transitions. Finite state machines model many electronic controllers and operating system command interpreters.

The typical artificial neural network is an even simpler model. It is a loop-free network of gates modeled after neurons. The gates are arranged in layers from those connected to inputs to those connected to outputs. A pattern of bits at the input passes through the network and produces an

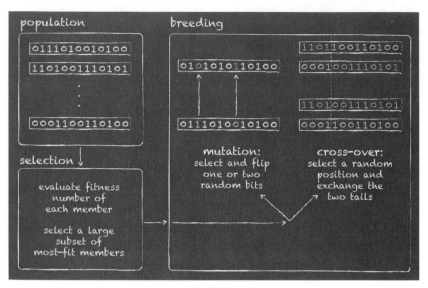

FIGURE 4. Since the 1950s, various geneticists have experimented with computer simulations of biological evolution, studying how various traits are passed on and how a population evolves to adapt to its circumstances. In 1975, John Holland adapted the idea of these simulations to a general method for finding near-optimal solutions to complex problems in many domains. The idea, depicted in the flow diagram above, is to develop a population of candidate solutions to the problem, encoded as bit-strings. Each bit-string is evaluated by a fitness function, and the most-fit members of the population are selected for reproduction by mutation and crossover. A bit-string is modified by mutation when one or several of its bits are randomly flipped. A pair of bit-strings is modified by crossover by selecting a random break point and exchanging the two tails of the strings. These changes generate a new population. The process is iterated many times until there are no further improvements in the most-fit individuals or until the computational budget is exhausted. This process is surprisingly good at finding near-optimal solutions to optimization problems whose direct solutions would otherwise be intractable.

output. There is no state to be recorded or remembered. Each signal from one layer to the next has an associated weight. The network is trained by an algorithm that iteratively adjusts the weights until the network becomes very good at generating the desired output. Some people call this *machine learning* because the trained (weight-adjusted) circuit acquires a capability to implement a function by being shown many examples. It is also called *deep learning* because of the hidden layers and weights in the circuit. Many modern advances in artificial intelligence and data analytics have been achieved by these circuits. Simulations of these circuits now allow for millions of nodes and dozens of layers.

When you go outside computer science, you will find few people talking about Turing machines and finite state machines. They talk instead of machine learning and simulation of computational models relevant to their fields. In each field, the computational designer either programs a model or designs a new model—or both.

An important issue with computational models is complexity—how long does it take to get a result? How much storage is needed? Very often a computational model that will give you the exact answer is impossible, too expensive, or too slow. Computational designers get around this with *heuristics*—fast approximations that generate close-approximation solutions quickly. Experimental validation is often the only way to gain trust in a heuristic. An artificial neural network for face recognition is a heuristic. No one knows of an exact algorithm for recognizing faces. But we know how to build a fast neural network that can get it right most of the time.

Advances and Limits

Computing has changed dramatically since the time when computational modeling grew up. In the 1980s, the hosting system for grand challenge models was a supercomputer. Today the hosting system is the entire Internet, now more commonly called "the cloud"—a massively distributed system of data and processing resources around the world. Commercial cloud services allow you to mobilize the storage and processing power you need when you need it. In addition, we are no longer constrained to deal with finite computations—those that start, compute, deliver their output, and stop. Instead we now tap endless flows of data and processing power as needed, and we count on the whole thing to keep operating indefinitely. With so much cheap, massive computing power, more people can be computational designers and tackle grand challenge problems.

But there are important limits to what we can do with all this computing power. One limit is that most of our computational methods have a sharp focus—they are very good at the particular task for which they were designed, but not for seemingly similar tasks. We can often overcome that limit with a new design that closes a gap in the old design. Facial recognition is an example. A decade ago, we did not have good methods of detecting and recognizing faces in images—we had to examine the images ourselves. Today, with deep learning algorithms, we have designed very reliable automated face recognizers, overcoming the earlier gap.

Another limit is that there are many problems that cannot be solved at all with computation. Some of these problems are purely technical, such as determining by inspection when a computer program will halt or enter an infinite loop. Many others are very complex issues, featuring technologies intertwined with social communities and no obvious answers—which are known as *wicked problems*. Many wicked problems are caused by the combined effects of billions of people using a technology. For example, the production of more than a billion refrigerators releases enough fluorocarbons to disrupt the upper atmosphere's protection against excessive sunlight. Millions of cars produce so much smog that some cities are unhealthy. The only solutions to these problems will emerge from social cooperation among the groups that now offer competing and conflicting approaches. Although computing technology can help by visualizing the large-scale effects of our individual actions, only social action will solve the problems we are causing.

Still, computational science is a powerful force within science. It emphasizes the "computational way" of doing science and turns its practitioners into skilled computational designers (and thinkers) in their fields of science. Computational designers spend much of their time inventing, programming, and validating computational models, which are abstract machines that solve problems or answer questions. Computational designers need to be computational thinkers as well as practitioners in their own fields. Computational design will be an important source of work in the future.

Bibliography

Aho, A. 2011. Computation and computational thinking. *Ubiquity* Symposium. DOI: 10.1145/1895419.1922682.

Baltimore, D. 2001. How biology became an information science. In *The Invisible Future: The Seamless Integration of Technology into Everyday Life*, ed. P. Denning, pp. 43–46. New York: McGraw-Hill.

Computer Science Teachers Association. 2011. Operational Definition of Computational Thinking for K-12 Education, https://www.csteachers.org/page/CompThinking.

Computing at School, a subdivision of the British Computer Society. 2015. *Computational thinking: A guide for teachers*. http://www.computingatschool.org.uk/computationalthinking.

Easton, T. 2006. Beyond the algorithmization of the sciences. *Communications of the ACM* 49(5):31–33.

Harvard Graduate School of Education. Computational thinking with Scratch: Defining. http://scratched.gse.harvard.edu/ct/defining.html.

Holland, J. 1975. *Adaptation in Natural and Artificial Systems*. Cambridge, MA: MIT Press.

Kelly, K. 2016. *The Inevitable: Understanding the 12 Technological Forces That Will Shape Our Future*. New York: Viking Books.

Papert, S. 1980. *Mindstorms: Children, Computers, and Powerful Ideas*. New York: Basic Books.

Tedre, M., and P. J. Denning. 2016. The long quest for computational thinking. *Proceedings of the 16th Koli Calling Conference on Computing Education Research*, November 24–27, 2016, Koli, Finland, pp. 120–129.

Wilson, K. G. 1989. Grand challenges to computational science. *Future Generation Computer Systems* 5:171–189.

Wing, J. 2006. Computational thinking. *Communications of the ACM* 49:33–35.

Quantum Questions Inspire New Math

Robbert Dijkgraaf

Mathematics might be more of an environmental science than we realize. Even though it is a search for eternal truths, many mathematical concepts trace their origins to everyday experience. Astrology and architecture inspired Egyptians and Babylonians to develop geometry. The study of mechanics during the scientific revolution of the seventeenth century brought us calculus.

Remarkably, ideas from quantum theory turn out to carry tremendous mathematical power as well, even though we have little daily experience dealing with elementary particles. The bizarre world of quantum theory—where things can seem to be in two places at the same time and are subject to the laws of probability—not only represents a more fundamental description of nature than what preceded it, it also provides a rich context for modern mathematics. Could the logical structure of quantum theory, once fully understood and absorbed, inspire a new realm of mathematics that might be called "quantum mathematics"?

There is of course a long-standing and intimate relationship between mathematics and physics. Galileo famously wrote about a book of nature waiting to be decoded: "Philosophy is written in this grand book, the universe, which stands continually open to our gaze. But the book cannot be understood unless one first learns to comprehend the language and read the letters in which it is composed. It is written in the language of mathematics."

From more modern times we can quote Richard Feynman, who was not known as a connoisseur of abstract mathematics: "To those who do not know mathematics it is difficult to get across a real feeling as to the beauty, the deepest beauty, of nature. . . . If you want to learn about nature, to appreciate nature, it is necessary to understand the language that she speaks in." (On the other hand, he also stated: "If all mathematics disappeared today, physics would be set back exactly one week," to which a mathematician had the clever riposte: "This was the week that God created the world.")

The mathematical physicist and Nobel laureate Eugene Wigner has written eloquently about the amazing ability of mathematics to describe reality, characterizing it as "the unreasonable effectiveness of mathematics in the natural sciences." The same mathematical concepts turn up in a wide range of contexts. But these days we seem to be witnessing the reverse: the unreasonable effectiveness of quantum theory in modern mathematics. Ideas that originate in particle physics have an uncanny tendency to appear in the most diverse mathematical fields. This is especially true for string theory. Its stimulating influence in mathematics will have a lasting and rewarding impact, whatever its final role in fundamental physics turns out to be. The number of disciplines that it touches is dizzying: analysis, geometry, algebra, topology, representation theory combinatorics, probability—the list goes on and on. One starts to feel sorry for the poor students who have to learn all this!

What could be the underlying reason for this unreasonable effectiveness of quantum theory? In my view, it is closely connected to the fact that in the quantum world everything that can happen does happen.

In a very schematic way, classical mechanics tries to compute how a particle travels from A to B. For example, the preferred path could be along a geodesic—a path of minimal length in a curved space. In quantum mechanics, one considers instead the collection of all possible paths from A to B, however long and convoluted. This is Feynman's famous "sum over histories" interpretation. The laws of physics will then assign to each path a certain weight that determines the probability that a particle will move along that particular trajectory. The classical solution that obeys Newton's laws is simply the most likely one among many. So, in a natural way, quantum physics studies the set of all paths, as a weighted ensemble, allowing us to sum over all possibilities.

This holistic approach of considering everything at once is very much in the spirit of modern mathematics, where the study of "categories" of objects focuses much more on the mutual relations than on any specific individual example. It is this bird's-eye view of quantum theory that brings out surprising new connections.

Quantum Calculators

A striking example of the magic of quantum theory is mirror symmetry—a truly astonishing equivalence of spaces that has revolutionized geometry. The story starts in enumerative geometry, a well-established, but not very exciting branch of algebraic geometry that counts objects. For example, researchers might want to count the number of curves on Calabi-Yau spaces—six-dimensional solutions of Einstein's equations of gravity that are of particular interest in string theory, where they are used to curl up extra space dimensions.

Just as you can wrap a rubber band around a cylinder multiple times, the curves on a Calabi-Yau space are classified by an integer, called the degree, that measures how often they wrap around. Finding the numbers of curves of a given degree is a famously hard problem, even for the simplest Calabi-Yau space, the so-called quintic. A classical result from the nineteenth century states that the number of lines—degree-one curves—is equal to 2,875. The number of degree-two curves was only computed around 1980 and turns out to be much larger: 609,250. But the number of curves of degree three required the help of string theorists.

Around 1990, a group of string theorists asked geometers to calculate this number. The geometers devised a complicated computer program and came back with an answer. But the string theorists suspected it was erroneous, which suggested a mistake in the code. Upon checking, the geometers confirmed that there was, but how did the physicists know?

String theorists had already been working to translate this geometric problem into a physical one. In doing so, they had developed a way to calculate the number of curves of any degree all at once. It's hard to overestimate the shock of this result in mathematical circles. It was a bit like devising a way to climb each and every mountain, no matter how high!

Within quantum theory, it makes perfect sense to combine the numbers of curves of all degrees into a single elegant function. Assembled in this way, it has a straightforward physical interpretation. It can be seen as a probability amplitude for a string propagating in the Calabi-Yau space, where the sum-over-histories principle has been applied. A string can be thought to probe all possible curves of every possible degree at the same time and is thus a superefficient "quantum calculator."

But a second ingredient was necessary to find the actual solution: an equivalent formulation of the physics using a so-called "mirror"

Calabi-Yau space. The term "mirror" is deceptively simple. In contrast to the way an ordinary mirror reflects an image, here the original space and its mirror are of very different shapes; they do not even have the same topology. But in the realm of quantum theory, they share many properties. In particular, the string propagation in both spaces turns out to be identical. The difficult computation on the original manifold translates into a much simpler expression on the mirror manifold, where it can be computed by a single integral. *Et voilà!*

Duality of Equals

Mirror symmetry illustrates a powerful property of quantum theory called duality: Two classical models can become equivalent when considered as quantum systems, as if a magic wand is waved and all the differences suddenly disappear. Dualities point to deep but often mysterious symmetries of the underlying quantum theory. In general, they are poorly understood and an indication that our understanding of quantum theory is incomplete at best.

The first and most famous example of such an equivalence is the well-known particle–wave duality, which states that every quantum particle, such as an electron, can be considered both as a particle and as a wave. Both points of views have their advantages, offering different perspectives on the same physical phenomenon. The "correct" point of view—particle or wave—is determined solely by the nature of the question, not by the nature of the electron. The two sides of mirror symmetry offer dual and equally valid perspectives on "quantum geometry."

Mathematics has the wonderful ability to connect different worlds. The most overlooked symbol in any equation is the humble equal sign. Ideas flow through it, as if the equal sign conducts the electric current that illuminates the "Aha!" lightbulb in our mind. And the double lines indicate that ideas can flow in both directions. Albert Einstein was an absolute master of finding equations that exemplify this property. Take $E = mc^2$, without a doubt the most famous equation in history. In all its understated elegance, it connects the physical concepts of mass and energy that were seen as totally distinct before the advent of relativity. Through Einstein's equation, we learn that mass can be transformed into energy, and vice versa. The equation of Einstein's general theory of relativity, although less catchy and well-known, links the worlds of geometry and matter in an equally surprising and beautiful manner. A succinct way to summarize that theory is that mass tells space how to curve, and space tells mass how to move.

Mirror symmetry is another perfect example of the power of the equal sign. It is capable of connecting two different mathematical worlds. One is the realm of *symplectic geometry*, the branch of mathematics that underlies much of mechanics. On the other side is the realm of *algebraic geometry*, the world of complex numbers. Quantum physics allows ideas to flow freely from one field to the other and provides an unexpected "grand unification" of these two mathematical disciplines.

It is comforting to see how mathematics has been able to absorb so much of the intuitive, often imprecise reasoning of quantum physics and string theory, and to transform many of these ideas into rigorous statements and proofs. Mathematicians are close to applying this exactitude to homological mirror symmetry, a program that vastly extends string theory's original idea of mirror symmetry. In a sense, they're writing a full dictionary of the objects that appear in the two separate mathematical worlds, including all the relations they satisfy. Remarkably, these proofs often do not follow the path that physical arguments had suggested. It is apparently not the role of mathematicians to clean up after physicists! On the contrary, in many cases completely new lines of thought had to be developed in order to find the proofs. This is further evidence of the deep and as yet undiscovered logic that underlies quantum theory and, ultimately, reality.

Niels Bohr was very fond of the notion of complementarity. The concept emerged from the fact that, as Werner Heisenberg proved with his uncertainty principle, in quantum mechanics one can measure either the momentum p of a particle or its position q, but not both at the same time. Wolfgang Pauli wittily summarized this duality in a letter to Heisenberg dated October 19, 1926, just a few weeks after the discovery: "One can see the world with the p-eye, and one can see it with the q-eye, but if one opens both eyes, then one becomes crazy."

In his later years, Bohr tried to push this idea into a much broader philosophy. One of his favorite complementary pairs was truth and clarity. Perhaps the pair of mathematical rigor and physical intuition should be added as another example of two mutually exclusive qualities. You can look at the world with a mathematical eye or with a complementary physical eye, but don't dare to open both.

Tangled Tangles

ERIK D. DEMAINE, MARTIN L. DEMAINE,
ADAM HESTERBERG, QUANQUAN LIU,
RON TAYLOR, AND RYUHEI UEHARA

The Tangle toy [14, 15] is a topological manipulation toy that can be twisted and turned in a variety of different ways, producing different geometric configurations. Some of these configurations lie in three-dimensional space, while others may be flattened into planar shapes. The toy consists of several curved, quarter-circle pieces fit together at rotational/twist *joints*. Each quarter-circle piece can be rotated about either of the two joints that connect it to its two neighboring pieces. Figure 1 shows a couple of Tangle toys that can be physically twisted into many three-dimensional configurations. See [15] for more information and for demonstrations of the toy.

More precisely, an n-Tangle consists of n quarter-circle *links* connected at n *joints* in a closed loop.[1] Tangles can be transformed into many configurations by rotating and/or twisting the joints along the axis of the two incident links (which must meet at a 180° angle). Figure 2 shows an example of such an axis of rotation along which joints may be rotated. The links connected to the joint in Figure 2 can be twisted clockwise or counterclockwise about the axis as shown by the arrows. While Tangle configurations usually lie in three-dimensional space, here we focus on *planar* Tangle configurations, or Tangle configurations that can be flattened on a flat surface.

Previous research into planar Tangle configurations by Chan [5] and Fleron [9] uses an analogy between Tangle and cell-growth problems involving polyominoes. An n-omino is composed of n squares of equal size such that every square is connected to the structure via incident edges. A well-known problem involving polyominoes is how many distinct *free* (cannot be transformed into each other via translations,

FIGURE 1. Two Tangle toys. Photo by Quanquan Liu, 2015. See also color image.

FIGURE 2. Blue dots represent joints. The axis is represented by the dotted line. Arrows show that both the red and black links can be rotated clockwise and counterclockwise about the axis. See also color image.

rotations, or reflections) n-ominoes there are for $n = 1, 2, 3, \ldots$. The number of free n-polyominoes up to $n = 28$ has been determined [10, 11, 12]. Figure 3 shows the two possible free configurations for the tromino.

Using the analogy to polyominoes, where a Tangle link represents a polyomino cell, Chan [5] and Fleron [9] pose two questions regarding planar Tangle configurations. First, what is the number of distinct planar n-Tangles for $n = 4i$, where $i = 1, 2, 3, \ldots$? (These numbers are known as the "Tanglegram sequence.") In other words, given a Tangle toy with n links, what is the number of distinct planar Tangles that can be formed? Second, can any planar n-Tangle be transformed into any other according to specified moves described in Section 2 of this

FIGURE 3. There are only two possible distinct free trominoes.

article? It has been conjectured, but not proven, that the answer to the second question is "yes."

The problem of determining whether all planar configurations can be reconfigured into each other using allowable moves is known as *flat-state connectivity* [4] of *linkages*. Recall that a Tangle toy is an example of a linkage that is composed of links and joints. One can think about the links as "edges" and the joints as "vertices"; just as vertices connect edges, joints connect links. Thus far, the study of flat-state connectivity has focused on *fixed-angle linkages*, where each link has an assigned fixed length and each vertex has an assigned fixed angle (i.e., the angle of incidence between two incident links is fixed). A *flat state* of such a linkage is an embedding of the linkage into $\mathbb{R}^2$. A linkage is *flat-state connected* if any two flat states of the linkage can be reconfigured into each other using a sequence of dihedral motions without self-intersections. Otherwise, the linkage is *flat-state disconnected*. All open chains with no acute angles, and all closed orthogonal unit chains, are flat-state connected, while open chains with $180°$ edge spins and graphs (as well as partially rigid trees) are flat-state disconnected [4]. For more details regarding these linkages, see Aloupis et al. [4]. Closed orthogonal unit fixed-angle chains (chains that have unit length edges and $90°$ angles of incidence between edges) move essentially like Tangles (viewing each quarter-circle link as a $90°$ corner between two half edges), so their flat-state connectivity means that there are (complex, three-dimensional) moves between any two planar Tangle configurations of the same length.

The previous study of reachable configurations of Tangles considered a set of moves called x- and Ω-rotations [5, 9]. We generalize these moves into two broad categories, reflections and translations, that allow for a larger set of possible moves. In particular, our reflection moves involve rotating one chain of the Tangle by $180°$ around the rest, effectively reflecting the former. Such reflections over an axis (such as "flipturns," "Erdős pocket flips," and "pivots") have been studied in previous work on transforming planar polygons [1–3, 8]. The purpose of such reflections is to simplify complex moves involving many edge flips and rotations into simpler, more "local" moves. The reflection and translation moves used here encompass all possible edge flips around any two joints in a Tangle; thus, they seem natural to use as simplifications of more complex Tangle moves. More details of these moves, as well as their relation to the previous x- and Ω-rotations, can be found in Section 3 of this article.

Our results show that Tangle configurations are flat-state disconnected under even our general reflection and translation moves, disproving a conjecture of Chan [5]. This result provides an example of nontrivial flat-state disconnectedness. Planar Tangle configurations are natural examples of flat-state configurations obtained using a set of local moves around two joints. We show examples of planar Tangle configurations that have no moves whatsoever, as well as examples that have a few moves but cannot escape a small neighborhood of configurations.

In addition to our results on Tangle flat-state connectivity, we present two different Tangle fonts. This is a continuation of a study on mathematical typefaces based on computational geometry, as surveyed in [6]. The two Tangle fonts were created from 52- and 56-Tangles.

We define our notation and conventions in Section 1. In Section 2, we describe the two classes of moves we considered in evaluating planar Tangles and their reachable configurations. In Section 3, we present examples of planar Tangles that are locked or rigid under our specified set of moves. In Section 4, we present the two Tangle fonts. Finally, in Section 5, we conclude with some open questions.

1 Definitions

A Tangle link can have two orientations with respect to the body of the structure: *convex* or *reflex*. These orientations are shown in Figure 4.

A *face* in a planar Tangle configuration consists of a set of convex links. Two faces are *tangential* if they are connected by reflex links. Figure 5 shows examples of faces. It has been shown that an n-Tangle can form planar Tangle configurations if and only if n is a multiple of 4 [9].

Using this definition of "face," we can further define the dual graph representation of a planar Tangle configuration to be a graph consisting of a vertex for each face of the configuration, with an edge connecting

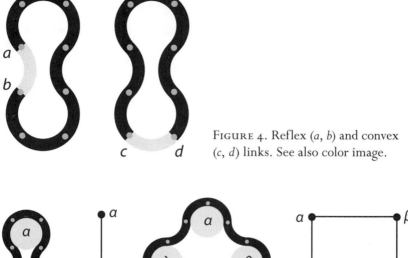

FIGURE 4. Reflex (*a*, *b*) and convex (*c*, *d*) links. See also color image.

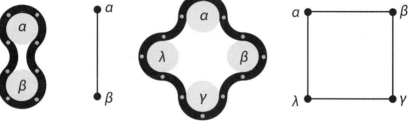

FIGURE 5. Dual graphs of planar Tangle configurations. The letters α, β, γ, and λ label faces that are connected to other faces by reflex links. See also color image.

each pair of tangential faces (Figure 5). This definition is analogous to the graph representation definition given by Taylor [13].

This *dual graph* representation of planar Tangle configurations is useful in certain proofs in the later sections. Furthermore, the dual graph representation is used by our planar Tangle moves enumerator to enable users to easily create arbitrarily shaped planar configurations [7].

2 Tangle Moves

We now describe the set of legal moves that can be performed on a planar Tangle configuration. We categorize these moves broadly as *translations* and *reflections*; the distinction is that translation moves are asymmetric, whereas reflection moves can be performed along a reflection axis.

2.1 Reflections

Reflection moves are performed over a linear *reflection axis*, which consists of a line through two joints of the Tangle. We call these two joints the *reflection joints*. To perform a reflection, one of the two parts of the Tangle separated by the reflection joints is rotated 180° clockwise or counterclockwise around the reflection axis. In fact, the previously mentioned x- and Ω-rotations [5] are reflection moves.[2]

The reflection move may only be made if the following two conditions hold:

1. There are no pieces occupying the space on the other side of the reflection.
2. The reflective joints are free to move 180° in the reflection direction (i.e., either clockwise or counterclockwise).

It not difficult to see when the second requirement is satisfied (specifically, when the reflection axis is exactly the axis of rotation of each of the joints). Figure 6 shows some examples of successful reflection moves.

A reflection around the axis can change the orientation of a link. For example, Figure 6 shows the result of reflecting a chain of links over the indicated x- or y-axis, resulting in the final configuration where all the orientations of the reflected links have changed. However, there are examples of reflections where the orientation of the links do not change (see the middle figure in Figure 6).

Some planar Tangle configurations may allow no reflection moves. Figure 7 shows two instances where no reflection moves are possible.

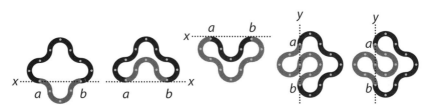

FIGURE 6. Reflection moves over horizontal and vertical axes where the joints labeled *a* and *b* are the reflection joints. Reflections may or may not change the orientation of the reflected links. See also color image.

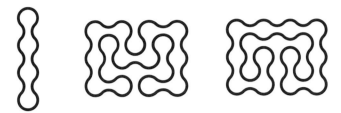

FIGURE 7. The first two planar Tangle configurations do not allow any reflection moves. The third configuration allows no translation moves.

2.2 TRANSLATIONS

Translations are asymmetric moves that are not performed across a single axis, but across a collection of parallel axes. A translation has two *translation joints*, oriented in the same (vertical or horizontal) direction, and four *translation links*, the two links next to each of the translation joints. When a translation move is performed, one of the two connected components of the Tangle without its translation links is picked up and translated to a different location relative to the other component by rotating the translation links. Figure 8 shows an example of a translation move.

Translation involves the rotation of the four links connected to the translational joints. A translation move may only be made if the following conditions hold:

1. The translational links can be rotated.
2. The translated portion may be placed in a location that does not contain other links.

The third configuration in Figure 7 shows an example where no translation moves are allowed.

The natural question (answered in Section 4) is whether any planar Tangle configurations can reach any other by a sequence of reflection and/or translation moves. If not, we call an *n*-Tangle *locked*, meaning that a proper subset of the planar configurations cannot reach configurations outside the set. In particular, we call a planar Tangle configuration *rigid* if it admits no such reflection or translation moves.

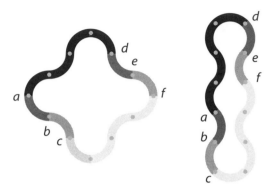

FIGURE 8. Example of a translation move. Translation involves the rotation of all four links connected to each of the two translational joints indicated by *b* and *e*. Here the translation move rotates the links spanned by the joints *a, b, c, d, e,* and *f.* See also color image.

2.3 TANGLE MOVES ENUMERATOR

The Tangle Moves Enumerator software [7] takes a starting planar Tangle configuration and lists all possible planar configurations that can be reached via the moves described in Sections 2.1 and 2.2. The enumerator performs the search in a breadth-first manner. There are $O(n^2)$ possible rotation and translation axes. For each axis, the number of possible rotational moves is two, and the number of possible translational moves is also two. Therefore, the number of possible new configurations resulting from moves in each level of the search is $O(n^2)$. The enumerator exhaustively searches each possible new configuration. If a configuration has been previously reached, the current branch of the search is terminated. In Section 3, we use this software to explore the configurations reachable from a planar Tangle configuration to determine whether it is locked or rigid.

3 Rigid and Locked Tangles

Here we illustrate two planar Tangle structures that are rigid under the moves defined in Sections 2.1 and 2.2. Furthermore, we demonstrate a set of locked but not rigid configurations with $n = 308$ links.

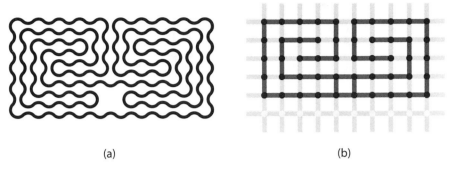

(a)   (b)

FIGURE 9. (a) A 4-leaf, symmetric, rigid counterexample. (b) The dual graph contains four leaves and a cycle. See also color image.

We thereby disprove conjectures by Chan [5] and Taylor [13]. Both examples can be verified by hand or with the Tangle Moves Enumerator (Section 2.3).

Figures 9 and 10 show two symmetric examples of rigid structures along with their dual graphs. In addition, Figure 10 shows that, even if we restrict ourselves to planar Tangle configurations where the dual graph is a path, there exist rigid configurations.

Figure 11 shows an example of locked, but not rigid, Tangles: these seven planar 308-Tangles cannot reach any planar configuration outside this set. Seven is far less than the number of possible planar 308-Tangle configurations, so the set is locked.

4 Tangle Fonts

Mathematical typefaces offer a way to illustrate mathematical theorems and open problems, especially in computational geometry, to the general public. Previous examples include typefaces illustrating hinged dissections, origami mazes, and fixed-angle linkages; see Demaine and Demaine [6]. Free software lets you interact with these fonts.[3]

Here we develop two Tangle typefaces, where each letter is a planar Tangle configuration of a common length. Figures 12 and 13 show the typeface of 52- and 56-Tangles, respectively. Our software allows you to write messages in these fonts.[4] We know that these configurations can reach each other by complex three-dimensional motions without

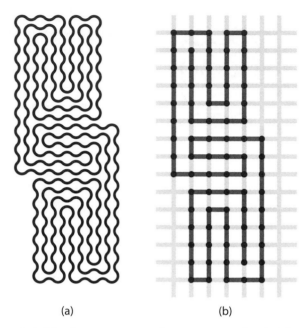

(a)   (b)

FIGURE 10. (a) A 2-leaf, symmetric rigid counterexample. (b) The dual graph contains two leaves and is a simple path. See also color image.

collision [2]. An interesting open question is whether the configurations in each font can reach any other using only reflection and translation moves. We conjecture that the answer is "yes"; see Figure 14 for one example.

5 Open Questions

Since we have shown that there exist planar locked and even rigid Tangle structures under our reflection and translation moves, a natural next step is to investigate the computational complexity of determining whether a structure is rigid. Another natural question is the computational complexity of determining whether two planar configurations of an n-Tangle can be transformed into each other through a sequence of valid moves. Furthermore, a natural optimization question is, given two planar n-Tangle configurations, find the minimal set(s) of reflection and translation moves necessary to transform one into the other.

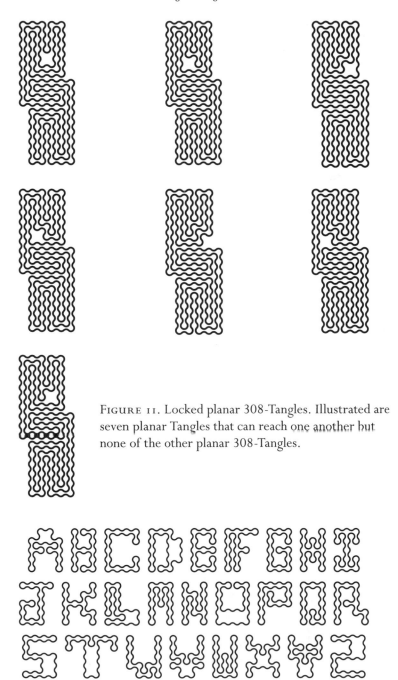

FIGURE 11. Locked planar 308-Tangles. Illustrated are seven planar Tangles that can reach one another but none of the other planar 308-Tangles.

FIGURE 12. 52-Tanglegram typeface.

FIGURE 13. 56-Tanglegram typeface.

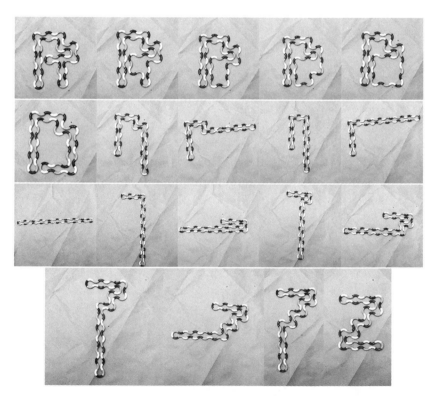

FIGURE 14. Transforming R into Z, in honor of Richard Zawitz, the inventor of Tangle [14].

Acknowledgment

We thank Julian Fleron and Philip Hotchkiss for inspiring initial discussions. We also thank the members of the MIT–Tufts Computational Geometry problem solving group for a fun and productive environment, and specifically Zachary Abel and Hugo Akitaya for helpful discussions related to Tangle.

Notes

1. Previous literature has also called the quarter-circles "pieces" or "segments." Here we choose to use "links" for greater specificity (when characterizing Tangle structures as fixed-angle linkages) and clarity.

2. Furthermore, the sequence of x-rotations introduced in [5] represents a type of translation move (described in Section 3.2 of [5]).

3. http://erikdemaine.org/fonts/.

4. http://erikdemaine.org/fonts/tangle/.

References

[1] O. Aichholzer, C. Cortés, E. D. Demaine, V. Dujmović, J. Erickson, H. Meijer, M. Overmars, B. Palop, S. Ramaswami, and G. T. Toussaint. Flipturning polygons. *Discrete Comp. Geometry* **28** no. 2 (2002) 231–253.

[2] O. Aichholzer, E. D. Demaine, J. Erickson, F. Hurtado, M. Overmars, M. Soss, and G. Toussaint. Reconfiguring convex polygons. *Comp. Geometry: Theory Applications*, **20** nos. 1–2 (2001) 85–95.

[3] G. Aloupis, B. Ballinger, P. Bose, M. Damian, E. D. Demaine, M. L. Demaine, R. Flatland, F. Hurtado, S. Langerman, J. O'Rourke, P. Taslakian, and G. Toussaint. Vertex pops and popturns. In *Proceedings of the 19th Canadian Conference on Computational Geometry (CCCG 2007)*. Ottawa, 2007, 137–140.

[4] G. Aloupis, E. D. Demaine, V. Dujmović, J. Erickson, S. Langerman, H. Meijer, J. O'Rourke, M. Overmars, M. Soss, I. Streinu, and G. Toussaint. Flat-state connectivity of linkages under dihedral motions. In *Proceedings of the 13th Annual International Symposium on Algorithms and Computation*. Vol. 2518 of *Lecture Notes in Computer Science*. Springer, New York, 2002, 369–380.

[5] K. Chan. Tangle series and tanglegrams: A new combinatorial puzzle. *J. Recreational Math.* **31** no. 1 (2003) 1–11.

[6] E. D. Demaine and M. L. Demaine. Fun with fonts: Algorithmic typography. *Theoretical Computer Science* **586** (June 2015) 111–119.

[7] E. D. Demaine and M. L. Demaine. Tanglegrams enumerator. Online at http://erikdemaine.org/tangle/, 2015 (last accessed June 29, 2016)

[8] E. D. Demaine, B. Gassend, J. O'Rourke, and G. T. Toussaint. All polygons flip finitely . . . right? In J. Goodman, J. Pach, and R. Pollack, editors, *Surveys on Discrete and Computational Geometry: Twenty Years Later*, 231–255. Vol. 453 of *Contemporary Mathematics*, American Mathematical Society, Providence, RI, 2008.

[9] J. F. Fleron. The geometry of model railroad tracks and the topology of tangles: Glimpses into the mathematics of knot theory via children's toys. Unpublished manuscript, February 2000.

[10] S. Mertens. Lattice animals: A fast enumeration algorithm and new perimeter polynomials. *J. Stat. Phys.* **58** no. 5 (1990) 1095–1108.

[11] T. Oliveira e Silva. Animal enumerations on the 4,4 euclidean tiling. Online at http://sweet.ua.pt/tos/animals/a44.html, 2015 (last accessed June 29, 2016).

[12] D. H. Redelmeijer. Counting polyominoes: Yet another attack. *Discrete Math.* **36**, no. 3 (1981) 191–203.

[13] R. Taylor. Planar tanglegrams. Presentation at MOVES 2015 conference, New York, August 2015, National Museum of Mathematics (MoMath).

[14] R. E. Zawitz. Annular support device with pivotal segments. United States Patent 4,509,929, April 9, 1985. Filed August 27, 1982.

[15] R. E. Zawitz. Tangle creations. Online at http://www.tanglecreations.com/, 2015, (last accessed June 29, 2016).

The Bizarre World of Nontransitive Dice: Games for Two or More Players

James Grime

Here is a game you can play with a friend. It is a game for two players with a set of three dice. These dice are not typical dice, however, because instead of having the values 1 to 6, they display various unusual values.

The game is simple: Each player picks a die. The two dice are then rolled together, and whoever gets the highest value wins.

The game seems fair enough. Yet, in a game of, say, ten rolls, you will always be able to pick a die with a better chance of winning—no matter which die your friend chooses. And you can make these dice at home right now.

Here is the set of three special dice (for figures with color labels in this article, also see color insert):

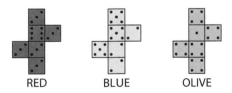

RED   BLUE   OLIVE

We say that A beats B if the probability of die A beating die B is greater than 50%.

It is simple to show that the Red die beats the Blue die by way of a tree diagram:

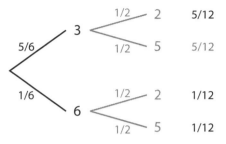

probability Red beats Blue = 7/12

From the diagram, we see that Red beats Blue with a probability of 7/12. This is greater than 50%, so Red is the better choice here.

Similarly, it can be shown that Blue beats Olive with a probability of 7/12. So we can set up a winning chain where Red beats Blue, and Blue beats Olive.

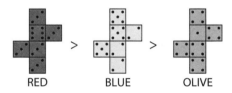

Using this information, it would be perfectly reasonable to expect, therefore, that Red beats Olive. If this is true, then we call the dice *transitive*.

However, this is not the case. In fact, bizarrely, Olive beats Red with a probability of 25/36. This means the winning chain is a circle, similar to the game Rock, Paper, Scissors.

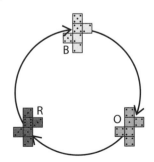

This is what makes the game so tricky because, as long as you let your opponent pick first, you will always be able to pick a die with a better chance of winning.

Double Whammy

After a few defeats, your friend may have become suspicious, but all is not lost. Once you explain how the dice beat each other in a circle, challenge your friend to one more game.

This time, you will choose first, in which case your opponent should be able to pick a die with a better chance of winning. But then increase both the stakes and the number of dice. This time, each player rolls two of his or her chosen die, and the player with the highest total wins.

Maybe using two dice means that your opponent has just doubled his or her chances of winning. But not so because, amazingly, with two dice the order of the chain flips!

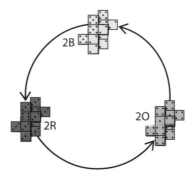

In other words, the chain reverses so that the circle of victory now becomes a circle of defeat, allowing you to win the game again!

Efron Dice

The paradoxical nature of nontransitive dice goes back to 1959 and to the Polish mathematicians Hugo Steinhaus and Stanislaw Trybuła [4], [5].

However, the remarkable reversing property is not true for all sets of nontransitive dice. For example, here is a set of four nontransitive dice introduced by Martin Gardner in 1970 [2]. This set was invented by the American statistician Brad Efron.

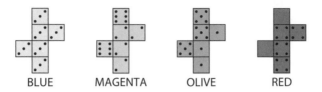

Here, the dice form a circle where Blue beats Magenta, Magenta beats Olive, Olive beats Red, and Red beats Blue, and they each do so with a probability of 2/3.

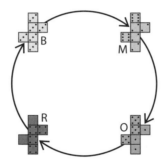

Trybula and Usiskin [6], [7] independently showed that one can always set up a nontransitive system of m n-sided dice and showed that the weakest winning probability has a bound. It is not possible for all winning probabilities to exceed this bound, but it is possible for all winning probabilities to be at least this bound, see Savage [3].

For six-sided dice, the set of three dice above achieve this bound. Using a different number of sides, the greatest bound for three dice is the golden ratio $\varphi = 0.618. \ldots$ This theoretical bound increases as the number of dice increases and converges to 3/4.

Efron dice achieve the bound for four dice of 2/3. Unfortunately, they do not possess the remarkable flipping property when you double the number of dice. Some of probabilities reverse, and some do not.

It is said that the billionaire American investor Warren Buffett is a fan of nontransitive dice. When he challenged his friend Bill Gates to a game, with a set of Efron dice, Bill became suspicious and insisted that Warren choose first. Maybe if Warren had chosen a set with a reversing property, he could have beaten Gates—he would just need to announce whether they were playing a one-die or two-dice version of the game after they had both chosen.

Three-Player Games

I wanted to know if it was possible to extend the idea of nontransitive dice to make a three-player game, i.e., a set of dice where two of your friends may pick a die each, then you can pick a die that has a better chance of beating both opponents at the same time!

It turns out that there is a way. The Dutch puzzle inventor M. Oskar van Deventer came up with a set of seven nontransitive dice with values from 1 to 21. Here, two opponents may each choose a die from the set of seven, and there will always be a third die with a better chance of beating each of them. The probabilities are remarkably symmetric; each arrow on the diagram illustrates a probability of 5/9.

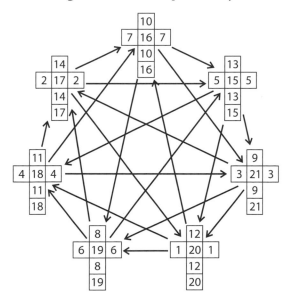

This means that we can play two games simultaneously; however, beating both players at the same time is still a challenge. The probability of doing so stands at around 39%.

This set of seven dice form a complete directed graph. In the same way, a four-player game would require 19 dice. A direct construction of this set was not known until 2016, when Angel and Davis devised a direct construction for any tournament of any number of dice [1].

However, I began to wonder if it was possible to exploit the reversing property of some nontransitive dice to design a slightly different three-player game, one that uses fewer than seven dice.

Grime Dice

My idea for a three-player game required a set of five dice that contained two nontransitive chains. When the dice were doubled, one chain would remain in the same order, while the second chain would reverse. This way, choosing a one-die or two-dice version of the game will allow you to play two opponents at the same time, no matter which dice they pick.

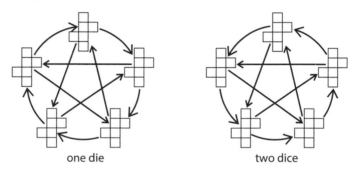

one die   two dice

After a small amount of trial and error, I devised the following set of five nontransitive dice.

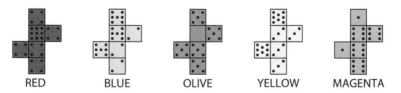

RED   BLUE   OLIVE   YELLOW   MAGENTA

These dice appeared to be the best set of five I could find. I have written about them before, and they became known as Grime dice.

For one-die games, we have the following chains:

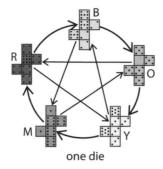

one die

All winning probabilities here are at least 5/9, with an average winning probability of 63%; I leave the calculations to the interested reader. Notice that the first chain is ordered alphabetically, whereas the second chain is ordered by word length of the color names.

You can also find nontransitive subsets of dice. For example, the Red, Blue, and Olive dice are a copy of the original set of three nontransitive dice that I describe above, complete with the same winning probabilities and reversing property.

For two-dice games, we get the following chains:

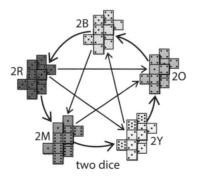

two dice

An unfortunate consequence of Red, Blue, and Olive having the reversing property is that, when we double the dice, the first chain (the outside circle) reverses order, while the second chain (the inside pentagram) stays the same—with one exception.

However, the probability of this exception is close to 50% (specifically, 625/1,296). Meanwhile, the average of all other winning probabilities is 62% (much higher than for Oskar dice), and so, in practice, the three-player game still works.

It is quite nice that this set of five dice contained three dice with their own reversing property. However, I admit, the exception continued to niggle at me. I wanted to know if there was a set of five nontransitive dice with the desired properties and no exceptions or if this set was really as close as we could get.

Finding a New Set of Grime Dice

I enlisted the help of a computer and the invaluable help of my friend Brian Pollock to search for sets of five nontransitive dice. The computational challenge of working out all sets of five dice and their chains was large one, so we devised a test.

Three dice can either form a diagram with all three arrows in the same direction, which we call a nontransitive chain, or with only two arrows in the same direction, which we call a transitive chain.

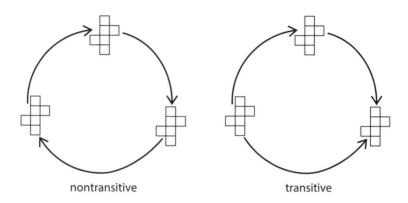

nontransitive   transitive

We wanted to create a set of five nontransitive dice, with two nontransitive chains, such that, when doubled, one chain stays the same and the other chain reverses order.

This means that, for any subset of three dice, if they form a nontransitive chain singly, then they will form a transitive chain when doubled. Alternatively, if they form a transitive chain singly, then they will form a nontransitive chain when doubled. If a chain remains transitive or nontransitive when the dice are doubled, then we say that the set has failed the test.

There are 10 subsets of three dice from a set of five. Each subset needs to pass the test. Furthermore, if all subsets pass the test, we have found a valid set of five dice with the desired properties.

Applying this test allowed us to reject sets without the desired property with less calculation.

Initially, we only considered dice using the values 0 to 9. Sets of dice that allow draws would be rather unsatisfactory. But after excluding draws, no set of five dice passed the test.

Only a few sets of four dice passed the test, which simply turned out to be the original Grime dice with one of the dice missing. This proved that Grime dice really are the best set of five dice using the values 0 to 9, without draws.

Dice with Higher Values

Naturally, the next thing to try were dice with higher values. Keeping the criterion of no draws, the first success found used the values 0 to 13.

A: 4, 4, 4, 4, 4, 9
B: 2, 2, 2, 7, 7, 12
C: 0, 5, 5, 5, 5, 10
D: 3, 3, 3, 3, 8, 13
E: 1, 1, 6, 6, 6, 11

There were two such sets using the values 0 to 13; the second set was only a slight variation of the above. These were also the only sets of five with the desired properties that use consecutive numbers.

I was delighted with this success, but the average winning probability is about 59%, lower than for Grime dice. So we continued our search to find a set with stronger winning probabilities.

The winning probabilities slowly increased as we increased the values on the dice. Here is one of the strongest sets of five dice using the values 0 to 17:

A: 4, 4, 8, 8, 8, 17
B: 2, 2, 2, 15, 15, 15
C: 0, 9, 9, 9, 9, 9
D: 3, 3, 3, 3, 16, 16
E: 1, 1, 10, 10, 10, 10

Increasing the dice values after this point did nothing to improve the winning probabilities. Since the numbers are no longer consecutive, there is enough space for the values to change without changing the winning probabilities, meaning that this set can appear repeatedly in slightly different forms. The investigation for better sets had plateaued.

For aesthetic reasons, I decided to subtract 8 from all sides of the above dice, making a set of new Grime dice (NGD) using values from −8 to 9:

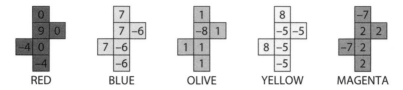

RED   BLUE   OLIVE   YELLOW   MAGENTA

Like the original Grime dice (OGD), this set makes two nontransitive chains, one with the colors listed alphabetically, the other with the colors listed by word length. When doubled, the alphabetical chain remains in the same order, whereas the chain ordered by word length flips.

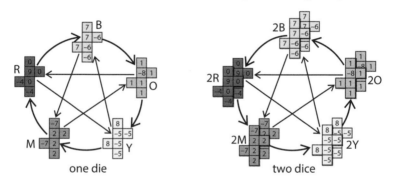

one die   two dice

In single dice games, NGD have the exact same winning probabilities as OGD. When the dice are doubled, NGD are generally slightly weaker, with average winning probability 60.4%, about 0.7% lower than for OGD. Crucially, however, all winning probabilities are now over 50%, allowing for a true three-player game as follows.

Invite two opponents to pick a die each, but do not volunteer whether you are playing a one-die or two-dice version of the game. No matter which dice your opponents pick, you will always be able to pick a die to beat each opponent. If your opponents pick two dice that are consecutive alphabetically, then play the one-die version of the game. If your opponents pick two dice that are consecutive by word length, then use the two-dice version of the game.

A Gambling Game

Can we expect to beat the two other players at the same time? Well, we have certainly improved the odds, with the average probability of beating both opponents now standing around 44%, a 5% improvement over Oskar dice. So, if the odds of beating two players is not over 50%, then how do we win? Consider the following gambling game.

Challenge two friends to a dice game where you will play your two opponents at the same time. If you lose, then you will give your opponent $1. If you win, then your opponent gives you $1. So, if you beat both players at the same time, then you win $2; if you lose to both players, then you lose $2, and if you beat one player but not the other, then your net loss is zero. You and your friends decide to play a game of 100 rolls.

If the dice were fair, then each player would expect to win zero since each player wins half the time and loses half the time.

However, with Oskar dice, you should expect to beat both players 39% of the time and lose to both players 28% of the time, which will give you a net profit of $22.

But even better, with new Grime dice, you should expect to beat both players 44.1% of the time but only lose to both players 23.6% of the time, giving you an average net profit closer to $41 (and possibly the loss of two former friends)!

I invite you to try out these games yourself and enjoy your successes and failures!

References

[1] L. Angel, M. Davis, A direct construction of non-transitive dice sets. *Journal of Combinatorial Designs* (2016) 10.1002/jcd.21563.

[2] M. Gardner, The paradox of the nontransitive dice and the elusive principle of indifference, *Sci. Amer.* **223** no. 6 (1970) 110–114.

[3] R. P. Savage Jr., The paradox of nontransitive dice, *Amer. Math. Monthly* **101** (1994) 429–436, http://dx.doi.org/10.2307/2974903.

[4] H. Steinhaus, S. Trybuła, On a paradox in applied probabilities, *Bull. Acad. Polon. Sci.* **7** (1959) 67–69.

[5] S. Trybuła, On the paradox of three random variables, *Zastos. Mat.* **5** (1960/1961) 331–332.

[6] ——, On the paradox of *n* random variables, *Zastos. Mat.* **8** (1965) 143–154.

[7] Z. Usiskin, Max-min probabilities in the voting paradox, *Annal. Mat. Stat.* **35** (June 1964), 857–862.

The Bingo Paradox

ARTHUR BENJAMIN, JOSEPH KISENWETHER, AND BEN WEISS

Believe it or not, when a large number of people play bingo, it is much more likely that the winning card has a completed horizontal row than a completed vertical column. How can this be? If you randomly mark off numbers on your own bingo card, you are just as likely to get a horizontal bingo as a vertical bingo. Why should the *winning* card be any different?

A computer program was written that generated 1,000 random bingo cards and played the game 100,000 times. To our surprise, horizontal winners were almost twice as likely as a vertical winner.

To better understand this paradox, let's review the rules of this popular game. A typical bingo card, like the one in Figure 1, has five columns, labeled B, I, N, G, and O, and each column has five numbers underneath it. Column B has five numbers from 1 through 15, in a random order. Similarly, columns I, N, G, and O have five random numbers from 16 to 30, 31 to 45, 46 to 60, and 61 to 75, respectively. Some bingo cards have a free space in the middle of column N.

The caller calls out numbers—such as "B11!"—pulled randomly from a container. Players place markers on the corresponding spaces on their cards if they have them. The first person to complete a row, column, or diagonal yells "Bingo!" and wins a prize. In the analysis that follows, we will initially ignore the effect of the free space, but we will consider it later.

Suppose that the first eight numbers were drawn in this order: B11, I23, G58, B13, I21, N34, G55, and O75. With these numbers, each letter has appeared, so it is possible for there to be a bingo card with a horizontal or diagonal win. But since no letter has appeared five times, it is impossible to have a vertical bingo at this point. This observation

FIGURE 1. A typical bingo card.

suggests a mathematical question that we can sink our teeth into. For a randomly generated sequence of bingo numbers, *what is the probability that all five letters appear before any single letter appears five times?*

Analyzing the Paradox

For simplicity, let's assume that we are playing with so many bingo cards that as soon as all five letters appear, we have a horizontal winner and as soon as one letter appears five times, we have a vertical winner.

We define a *horizontal sequence* to be an arrangement of the 75 bingo numbers so that all five letters appear before any letter appears five times. Otherwise, it is called a *vertical sequence*. We say a sequence has property H_n if it becomes a horizontal sequence on the nth number and has property V_n if it becomes a vertical sequence on the nth number. A sequence that begins with the eight numbers given above would have property H_8. Note that it is impossible for a sequence to be both horizontal and vertical since a number like O75 could not simultaneously be the first time and the fifth time that the letter O appears in the sequence.

The probability of a horizontal sequence in five draws (the minimum number possible) is

$$P(H_5) = \frac{60}{74} \cdot \frac{45}{73} \cdot \frac{30}{72} \cdot \frac{15}{71} \approx 0.04400$$

since after the first number is drawn, 60 of the remaining 74 numbers will produce a new letter, then 45 of the remaining 73 numbers will produce a third letter, and so on. Similarly, the chance of a vertical sequence in five draws is

$$P(V_5) = \frac{11}{74} \cdot \frac{13}{73} \cdot \frac{12}{72} \cdot \frac{11}{71} \approx 0.00087$$

So a horizontal sequence is about 50 times more likely than a vertical sequence on the fifth draw.

What about after the fifth draw? Here, the counting gets a little more complicated, but we can do it. Let's find the probability of achieving a horizontal sequence in exactly 10 draws. Note that there are 75! equally likely sequences of bingo numbers. How many of them result in a horizontal sequence on the tenth draw? Let's choose the tenth number first. There are 75 possibilities, so let's mentally choose the tenth to be O75.

The nine previous numbers can have four possible *shapes* based on how many of each of the letters B, I, N, and G appear: 4311, 4221, 3321, or 3222. For example, the sequence N31, N41, G59, I26, B5, N35, B8, B9, B7 (inspired by the digits of pi) has shape 4311, since a letter appears four times (B), another letter appears three times (N), and the other letters (I and G) each appear once.

The horizontal shapes are *partitions* of the integer 9 into four positive parts, where all parts have size at most 4. Thus, a shape like 5211 is not allowed since it contains five of the same letter and would therefore be a vertical sequence.

The letters B, I, N, and G can be given a shape of 4311 in 12 different ways (BBBBIIING, BBBBINNNG, and so on). Likewise, there are 12 ways to give them a shape of 4221, 12 ways to give them a shape of 3321, but only four ways to give them a shape of 3222 (we have four choices for which letter appears three times, and the other letters must all appear twice). In general, a four-digit shape can be assigned to B, I, N, and G in 1, 4, 6, 12, or 24 ways, depending on whether it consists of all the same digit, a tripled digit, two pairs of digits, one pair of digits, or all different digits, respectively.

Once we have determined how many of each letter is to be used, we can count the ways to assign them numbers using binomial coefficients. Recall that the binomial coefficient

$$\binom{15}{k} = \frac{15!}{k!(15-k)!}$$

is the number of ways we can choose k items out of a collection of 15 items. For example, the number of ways to assign numbers to BBBBIIING is

$$\binom{15}{4}\binom{15}{3}\binom{15}{1}\binom{15}{1} = 1{,}365 \cdot 455 \cdot 15^2 = 139{,}741{,}875$$

Say we choose the numbers B1, B2, B3, B4, I16, I17, I18, N31, G46. Then these nine numbers can be arranged, like Scrabble tiles in a rack, in 9! ways. Finally, the remaining 65 numbers (appearing after O75) can be arranged in 65! ways. Putting this all together, the probability that a bingo sequence becomes horizontal on the tenth draw with shape 4311 is

$$P(4311) = \binom{15}{4}\binom{15}{3}\binom{15}{1}\binom{15}{1}\frac{75 \cdot 12 \cdot 9!65!}{75!} \approx 0.01517$$

Similarly, the probabilities of becoming horizontal with the other possible shapes are

$$P(4221) = \binom{15}{4}\binom{15}{2}\binom{15}{2}\binom{15}{1}\frac{75 \cdot 12 \cdot 9!65!}{75!} \approx 0.02451$$

$$P(3321) = \binom{15}{3}\binom{15}{3}\binom{15}{2}\binom{15}{1}\frac{75 \cdot 12 \cdot 9!65!}{75!} \approx 0.03540$$

and

$$P(3322) = \binom{15}{3}\binom{15}{2}\binom{15}{2}\binom{15}{2}\frac{75 \cdot 4 \cdot 9!65!}{75!} \approx 0.01906$$

Altogether, the probability of achieving a horizontal sequence on the tenth draw is

$$P(H_{10}) = P(4311) + P(4221) + P(3321) + P(3222) \approx 0.09415$$

We perform similar calculations to find $P(V_{10})$, the probability of a vertical sequence on the tenth draw. Again, there are 75 possibilities for the tenth draw, which we will assume is O75. The previous nine numbers must include exactly four Os, which can be chosen in $\binom{14}{4}$ ways. The remaining five letters can have one of four possible shapes: 4100, 3200, 3110, or 2210. We exclude the shapes 5000 and 2111, since the first shape would create an earlier vertical sequence and the second shape (combined with the other Os) would create an earlier horizontal sequence. Each of the shapes has one digit that appears twice and can therefore be assigned letters in 12 ways. Therefore,

$$P(4100) = \binom{14}{4}\binom{15}{4}\binom{15}{1}\frac{75 \cdot 12 \cdot 9!65!}{75!} \approx 0.00223$$

$$P(3200) = \binom{14}{4}\binom{15}{3}\binom{15}{2}\frac{75 \cdot 12 \cdot 9!65!}{75!} \approx 0.00519$$

$$P(3110) = \binom{14}{4}\binom{15}{3}\binom{15}{1}\binom{15}{1}\frac{75 \cdot 12 \cdot 9!65!}{75!} \approx 0.01113$$

and

$$P(2210) = \binom{14}{4}\binom{15}{2}\binom{15}{2}\binom{15}{1}\frac{75 \cdot 12 \cdot 9!65!}{75!} \approx 0.01797$$

So, the probability of achieving a vertical sequence on the tenth draw is

$$P(V_{10}) = P(4100) + P(3200) + P(3110) + P(2210) \approx 0.03652$$

Comparing $P(V_{10})$ to $P(H_{10})$, we see that even on the tenth draw, horizontal sequences are more than twice as likely to appear as vertical sequences.

Every sequence will become horizontal or vertical within 17 draws: After 16 draws, we can have a sequence of four Bs, Is, Ns, and Gs, say, but the next number will be either an O (creating a horizontal sequence) or a B, I, N, or G (creating a vertical sequence). We summarize our findings in Table 1.

TABLE 1. The Probabilities of a Vertical or a Horizontal Bingo

n	Shapes	$P(H_n)$	$P(V_n)$	Ratio	Sum	Cumulative
5	1	0.04400	0.00087	50.57	0.0449	0.0449
6	1	0.08800	0.00373	23.60	0.0917	0.1366
7	2	0.11350	0.00956	11.87	0.1231	0.2597
8	3	0.12052	0.01903	6.33	0.1396	0.3992
9	4	0.11220	0.02902	3.87	0.1412	0.5404
10	4	0.09415	0.03652	2.58	0.1307	0.6711
11	5	0.07191	0.03968	1.81	0.1116	0.7827
12	4	0.04972	0.03748	1.33	0.0872	0.8699
13	4	0.03075	0.03085	1.00	0.0616	0.9315
14	3	0.01658	0.02178	0.76	0.0384	0.09698
15	2	0.00742	0.01260	0.59	0.0200	0.9898
16	1	0.00254	0.00558	0.45	0.0081	0.9980
17	1	0.00052	0.00151	0.34	0.0020	1.0000
Total	**35**	**0.752**	**0.248**		**1.000**	

The upshot is that the probability of a horizontal sequence is 75.2 percent, which is about three times more likely than a vertical sequence. A sequence becomes horizontal or vertical by the twelfth draw about 87 percent of the time. In all these cases, horizontal sequences are much more likely than vertical sequences. When it happens on the thirteenth draw (about 6 percent of the time) the sequences have almost the same probability, and when it happens after the thirteenth draw, which is only about 7 percent of the time, then the vertical sequences have the edge.

When all cards begin with a free space in the middle, the chance of a vertical sequence increases slightly, since column N now has 15 numbers instead of 14 numbers to cover the remaining four spaces. Joe Kisenwether and Dick Hess independently discovered that when the free space is used, the chance of a horizontal win is 73.73 percent (see Dick Hess's *The Population Explosion and Other Mathematical Puzzles*, World Scientific, 2016).

The Numbers of Shapes

Although we have answered our original question, more interesting mathematics is lurking behind the analysis.

When we enumerated sequences with properties H_{10} and V_{10}, we had to analyze—in both cases—exactly four shapes. This is not a coincidence. For each n, the sequences that are horizontal and vertical on the nth draw yield the same number of shapes. This result is not obvious. The number of shapes for H_n is the number of partitions of $n - 1$ by four positive integers less than 5, and the number of shapes for V_n is the number of partitions of $n - 5$ by four nonnegative integers less than 5, at least one of which is 0. What's going on?

We will illustrate the one-to-one correspondence between the shapes for H_n and the shapes for V_n in the case $n = 10$, but the reasoning is the same for any n.

The shapes used for the enumeration of H_{10} consist of four positive numbers that add to 9, where all numbers are less than or equal to 4. These partitions, 4311, 4221, 3321, and 3222, are displayed pictorially in Figure 2. Such representations are called *Ferrers diagrams*. Note that they fit in a 4-by-4 box.

If we subtract 1 from each value, or equivalently, delete the first columns of dots in the Ferrers diagrams, as in Figure 3, we get partitions

FIGURE 2. Ferrers diagrams for the partitions 4311, 4221, 3321, and 3222 (left to right) fit into a 4-by-4 box. See also color image.

FIGURE 3. Partitions of 5 into nonnegative values less than or equal to 3 fit into a 4-by-3 box. See also color image.

FIGURE 4. The conjugate partitions of those in Figure 3 fit into a 3-by-4 box. See also color image.

of 5 into four nonnegative parts, where all numbers are less than or equal to 3. The partitions now fit inside a 4-by-3 box.

Next, interchange the rows and columns of dots to create the *conjugate partitions* in Figure 4. These Ferrers diagrams fit into a 3-by-4 box, or equivalently a 4-by-4 box with an empty last row. They correspond to partitions of 5 by four nonnegative integers less than or equal to 4, at least one of which is 0. In particular, these are the V_{10} shapes 2210, 3110, 3200, and 4100.

Thus, this technique gives a bijection between the shapes for H_{10} and the shapes for V_{10}; and the same technique works for other values of n.

In fact, we can get more information from these diagrams. In the V_n case, each partition carves out a *lattice path* from the point $(0,0)$ to $(4,3)$ in the 3-by-4 box. For example, the partition 3200 creates the lattice path in Figure 5. Consequently, the number of shapes needed to

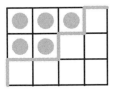

FIGURE 5. The lattice path corresponding to the partition 3200. See also color image.

compute all of the vertical sequences (and hence the number needed to compute all of the horizontal sequences) corresponds to the number of lattice paths from (0,0) to (4,3). Since each lattice path takes, in some order, four steps to the right and three steps up, the total number of shapes is

$$\binom{7}{3} = \frac{7!}{3!4!} = 35$$

as seen at the bottom of the second column of Table 1.

Finally, there is a slick way to generate the number of shapes for each n (the rest of the entries in the second column of Table 1) using *q-binomial coefficients*, which are polynomial generalizations of binomial coefficients.

First we replace the integer m with the polynomial

$$m_q = 1 + q + q^2 + \cdots + q^{m-1} = \frac{1-q^m}{1-q}$$

For example,

$$7_q = 1 + q + q^2 + \cdots + q^6 = \frac{1-q^7}{1-q}$$

Then a binomial coefficient like

$$\binom{7}{3} = \frac{7 \cdot 6 \cdot 5}{3 \cdot 2 \cdot 1} = 35$$

has a corresponding q-binomial coefficient

$$\binom{7}{3}_q = \frac{7_q \cdot 6_q \cdot 5_q}{3_q \cdot 2_q \cdot 1_q} = \frac{(1-q^7)(1-q^6)(1-q^5)}{(1-q^3)(1-q^2)(1-q^1)}$$

$$= \frac{1-q^5-q^6-q^7+q^{11}+q^{12}+q^{13}-q^{18}}{1-q-q^2+q^4+q^5-q^6}$$

Believe it or not, after simplifying this rational function, we obtain a twelfth-degree polynomial in which the coefficient of the q^n term is the number of partitions of the integer n that fit in a 4-by-3 box (see *Integer Partitions* by George E. Andrews and Kimmo Ericsson, Cambridge University Press, 2004, for a justification). In other words, it's the shape-counting polynomial

$$1 + q + 2q^2 + 3q^3 + 4q^4 + 4q^5 + 5q^6 + 4q^7 + 4q^8 + 3q^9 + 2q^{10} + q^{11} + q^{12}$$

as seen in the second column of Table 1.

The Sleeping Beauty Controversy

Peter Winkler

1. Introduction

Consider the following problem, stated here in the third person (Sleeping Beauty, or SB):

> Sleeping Beauty agrees to the following experiment. On Sunday, she is put to sleep, and a fair coin is flipped. If it comes up Heads, she is awakened on Monday morning; if Tails, she is awakened on Monday morning and again on Tuesday morning. In all cases, she is not told the day of the week, is put back to sleep shortly after, and will have no memory of any Monday or Tuesday awakenings.
>
> When Sleeping Beauty is awakened on Monday or Tuesday, what—to her—is the probability that the coin came up Heads?

When philosopher Adam Elga brought this puzzle to the world in 2000, he asked about degree of belief rather than probability; I use the latter term since mathematicians are accustomed to assigning numerical values to that quantity, as computed or estimated by a particular person about a particular event in a particular circumstance.

The issue is not that the Sleeping Beauty problem is undecidable (that is, not solvable from the axioms of set theory); indeed, it seems that nearly everyone discussing the problem has a strong opinion about its answer. The arguers fall into camps and subcamps, some claiming that the answer comes down to the nature of probability, the meaning of consciousness, evidential versus causal decision theory, one-world versus many-world quantum mechanics, conditioning versus updating—or exactly how the problem is phrased.

Those who believe that the answer is 1/2 have been dubbed "halfers," while those who go with 1/3 are called "thirders." If you believe either answer could be correct—depending, perhaps, on interpretation or phrasing—you are a "dualist"; if you think there is no correct answer because the problem cannot be well posed, you'll be called an "objector." There are subcamps (see below), and a few folks who say they simply don't know. Perhaps there should even be a category for those who don't care. But, hey, you've read this far, right?

Below, I will present the main arguments of the halfers and thirders, temporarily omitting references in order to maintain brevity. Then I will review the problem's history and go back over the arguments in reverse order, looking for enlightenment.

INFORMATION. Say the halfers: Before SB is put to sleep on Sunday, her credence that the fair coin will come up Heads is inarguably 1/2. She knows she will be awakened, so when she inevitably is, *she has no new information*, and therefore, her credence in Heads cannot have changed.

REFLECTION. Similarly: On Sunday, the thirder SB *knows* that, on Monday, she will give credence 1/3 to Heads. But then she should already have credence 1/3 on Sunday, which is absurd.

REPETITION. Say the thirders: Repeat the experiment 100 times; then SB will be awakened about 150 times, 50 of them to Heads. So SB's probability of Heads, on awakening, must be 1/3.

GAMBLING. Say the thirders: Ask SB, upon each awakening, if she's willing to have $3 deducted from her bank account if the coin landed Heads, provided that $2 is *added* to her account if the coin landed Tails. (She understands that if it's Tuesday morning and she accepted the bet on Monday, her holdings may already have changed—but that should have no influence on today's decision.)

As a thirder, SB should accept the bet: Her expectation is $1/3\ (-\$3) + 2/3\ (+\$2) > 0$. And she's right to do so: Over the course of the experiment, she ends up $4 ahead if the coin landed Tails and only $3 behind if it landed Heads. A corresponding calculation by the halfer SB would lead her to refuse the bet.

SYMMETRY. Suppose that 15 minutes after her Monday awakening, SB will be told what day it is. If she hears that it's Monday, her probability of Heads is 1/2. Suppose instead that SB will be told the result of the coin flip. Then, if it's Tails, her probability that it's Monday is 1/2. The first implies $\mathbb{P}$(Heads and Monday) $=\mathbb{P}$(Tails and Monday); the second, that $\mathbb{P}$(Tails and Monday) $=\mathbb{P}$(Tails and Tuesday). These three probabilities represent exhaustive and mutually exclusive events, thus each is equal to 1/3.

2. History

The above symmetry argument appeared in Elga's seminal paper [12]. Elga extracted the Sleeping Beauty problem (named by Robert Stalnaker) from Example 5 of Michele Piccione et al. [31], one of many papers in a volume of *Games and Economic Behavior* dedicated to a decision-theoretic problem known as "The Absent-Minded Driver."[1] Thus began a storm of arguments, papers, and blog comments, drawing in philosophers, mathematicians, and even physicists, then (seemingly) everyone.

Philosophers are after much bigger game than the Sleeping Beauty problem itself. How does one determine the credence that should be given to a particular proposition in given circumstances, and how should that credence be updated with new information or the passage of time? Sleeping Beauty is a demanding test for any theory that addresses these questions, incorporating loss of consciousness, loss of memory, and absence of time indication. Many philosophers acknowledge the success of Kolmogorov's probability axioms [24] but question whether they are equipped for handling propositions like "I am thirsty" whose truth-values change with time and perspective.

The debate in the philosophy community is thus concerned not just with the "correct" answer to Sleeping Beauty, but with which (if any) answer is implied by a given theory. Some, like Michael Titelbaum (see, e.g., [41, 42, 43] but not [40]) doubt that there is a good reason to choose any current framework over the rest in an effort to settle the issue. But even if you agree with this sentiment, you might ask which answer you *want* your theory to provide, and here the evidence (to me) is that the thirder position has emerged as the dominant view. This conclusion is not arrived at by counting papers as if they were votes since, of course, papers are supposed to present new ideas. Many published papers are attacks on the thirder position from various other camps, many others rebuttals. But anti-thirders seem to see themselves as fighting the establishment; concedes halfer Joel Pust [34]: "Most of those writing on the SB problem have argued that one-third is the correct answer."

First to challenge Elga was David Lewis [25], rebutted by Cian Dorr [10]. Frank Arntzenius started as an objector [1] but became a thirder [2]. Rachel Briggs [5] advanced the idea that causal[2] decision theorists (like her) should be thirders, and evidential decision theorists should be halfers, refuted by Vincent Conitzer [7]. Peter J. Lewis [26] argued that the many-world view of quantum physics implies halfism, rebutted by Alistair Wilson [47]. John Pittard, a halfer himself, argues in [33] that halfers must endorse robust perspectivalism.[3] More challenges and rebuttals are cited below in connection with particular arguments.

There have been many attempts to reconcile thirders and halfers by saying both are right (see Jacob Ross [36], who reaches a "rational dilemma," or Berry Groisman [15]), or both wrong (Nick Bostrom [4]); there is even (see Namjoong Kim [23]) a camp (the "lessers") who hold that the answer is less than $1/2$ but maybe not $1/3$. Jessi Cisewski et al. [6] claim that both sides, and all values between, are supportable, depending on whether one believes that SB's total knowledge—including degree of indigestion—is assumed to be exactly the same Monday and Tuesday.[4] Pradeep Mutalik [30], who writes for the excellent online magazine *QUANTA*, believes the answer depends on whether the question asks about "the coin associated with this experiment" or "the coin associated with this awakening." But most mathematicians would not, I think, accept the notion that equivalent events can have different probabilities.

To see why the thirders are leading, I will revisit Elga's symmetry argument, which looks a lot like a proof of the thirder position. Is it? We mathematicians like to think that an argument either is or isn't a proof, but this applies only to formal proofs (which, like honest politicians, are much talked about but rarely seen). In real life, much more is demanded of proofs of theorems that have counterintuitive consequences. I will therefore return in reverse order to the earlier arguments, ending with the halfers' compelling information argument. Perhaps the consequences of the thirder position can be made more comfortable.

3. Revisits

SYMMETRY REVISITED. Elga's symmetry argument retains the honor of being the most popular target of nonthirders and has been attacked on every conceivable front. Here is Elga's argument in symbolic form.

At a moment of SB's awakening there are three possible relevant states: Heads (H) and Monday (M); Tails (T) and Tuesday (U); T and M. If SB is to be told (on every awakening) the day of the week and is now told "Monday," then, by coin-flip symmetry,

$$\mathbb{P}(H\,|\,M) = \mathbb{P}(T\,|\,M)$$

If SB is to be told (on every awakening) the state of the coin and is now told "Tails," then, this time by indistinguishability of the Monday and Tuesday awakenings,

$$\mathbb{P}(M\,|\,T) = \mathbb{P}(U\,|\,T)$$

Therefore,

$$\mathbb{P}(H \wedge M) = \mathbb{P}(H\,|\,M) \cdot \mathbb{P}(M) = \mathbb{P}(T\,|\,M) \cdot \mathbb{P}(M) = \mathbb{P}(T \wedge M)$$

and

$$\mathbb{P}(T \wedge M) = \mathbb{P}(M\,|\,T) \cdot \mathbb{P}(T) = \mathbb{P}(U\,|\,T) \cdot \mathrm{P}(T) = \mathbb{P}(T \wedge U)$$

so

$$\mathbb{P}(H \wedge M) = \mathbb{P}(T \wedge M) = \mathbb{P}(T \wedge U)$$

Since these events are mutually exclusive and exhaustive, their probabilities sum to 1, so each has probability $1/3$ and in particular

$$\mathbb{P}(H) = \mathbb{P}(H \wedge M) = 1/3$$

Some halfers, including Lewis [25], dispute the claim that $\mathbb{P}(H\,|\,M) = 1/2$—a hard position to maintain when you consider that the experimenters don't need to flip the coin until Monday night. (How can SB's credence in Heads be other than $1/2$ if she knows the coin hasn't been flipped yet?) Others (called "double-halfers") concede that $\mathbb{P}(H\,|\,M) = 1/2$, maintaining that SB's credence in Heads doesn't change when she hears that it's Monday. How, then, do they deal with the math? Mikaël Cozic [9] suggests that conditioning is not the right way to modify SB's credence; Ioannis Mariolis [27] claims there are two kinds of "it is Monday" events, one of which he calls "Monday*"; Joseph Halpern [16] claims a "difference between the probability of heads conditional on it being Monday versus the probability of heads conditional on learning that it is Monday";[5] Patrick Hawley [17] does not accept that SB should be uncertain of the day![6] Roger White [46] doesn't know what's wrong with Elga's argument but claims that a natural generalization of it has an unacceptable consequence. (Terry Horgan [20] is happy with the generalization but argues that its consequence is not only acceptable but demonstrably correct.) As far as I can tell, every published attack on Elga's argument has been riposted, except perhaps for the just-published [6], whose quarrel with Elga is that he *assumes* that when SB is awakened, the events "it is Monday" and "it is Tuesday" are exclusive and exhaustive.

To help clarify the main camps, here's a rephrasing (some might disagree) of Sleeping Beauty: Alice, Bob, and Charlie have each taken a new sleeping pill. In hospital experiments, half the subjects slept through the night (as intended by the pill's creators), but the other half woke up once in the middle of the night then returned to sleep and woke up in the morning with no memory of the night awakening.

Alice wakes up in the middle of the night, and her credence that the pill has worked drops to zero (no argument there).

Bob wakes up in the morning; his credence that the pill worked remains at $1/2$ (thirders and double-halfers would agree, but the Lewisians would not).

Charlie, who has blackout shades in his bedroom, wakes up not knowing whether it is morning. According to the thirders, his credence in the efficacy of the pill is $1/3$ until he raises the shades, at which point it rises to $1/2$ or drops to zero. If he desperately wants the pill to have worked, you might think Charlie would be happy to see the morning sun—but would he? The double-halfers believe that the sun does not change Charlie's credence that the pill worked.

Further symmetry arguments have been advanced. Modulo details, Dorr [10] and Arntzenius [2] suppose that regardless of the coin flip, SB will be awakened on both days, but in the event of Heads, SB will be told "Heads and Tuesday" 15 minutes after her Tuesday awakening. Then, for the first 15 minutes that SB is awake, "Heads and Monday," "Heads and Tuesday," "Tails and Monday," and "Tails and Tuesday" are equiprobable by symmetry. After 15 minutes, when SB doesn't get the "Heads and Tuesday" signal, she eliminates that option, and the other three possibilities must remain equiprobable.

Other mathematical arguments, including those in Jeffrey Rosenthal's [35] in *The Mathematical Intelligencer* and Titelbaum's 2008 paper [41], have been offered. These as well are plausible to a mathematician—except, of course, for possible unintuitive consequences of the thirder position, which will be addressed below.

GAMBLING REVISITED. Both halfers and thirders have attempted to employ "Dutch books" to discredit the opposition; some of their arguments can be found in Cristopher Hitchcock [18] (great title). A Dutch book is a sequence of "fair" bets with a guaranteed negative outcome—of course, this ought not to be possible, so the argument is that, if a Dutch book can be made, the probabilities upon which the fairness of the bets relies must not be correct. (Yes, regardless of SB's credences, she can always work out her best strategy for the whole experiment and bet accordingly. But her credences ought to be reliable guides.)

Some Dutch book (and other decision-theoretic) arguments present situations in which SB's best choice at a given awakening might depend on her decision at a previous or forthcoming awakening. In such cases, causal and evidential decision theory might differ concerning her rational action. A causal decision theorist believes SB's actions should be based only on what they cause (in particular, *not* on her decisions at other awakenings), whereas an evidential decision theorist is permitted to use, e.g., her judgment that her decisions at all awakenings will probably be the same. Readers familiar with the Prisoner's Dilemma (see, e.g., Steven Tadelis [39]) might imagine a case where the prisoners are identical twins who don't necessarily give a fig for one another but have historically always made the same decisions in identical situations. If the twins are evidential decision theorists, they will escape the dilemma, each reasoning that whatever he does, his twin will probably do the same. If they are causal decision theorists, too bad!

Conitzer [8] regards as suspect any decision-theoretic argument in which SB's decisions are not "additive," therefore independent of decisions at other awakenings. Conitzer shows that the thirder SB plays additive games optimally but does not believe that this settles the SB problem conclusively.

REPETITION REVISITED. To the argument that 1/3 of awakenings are Heads, the halfers reply that one should count experiments—that is, coin flips—not awakenings. Being asked twice upon Tails, they say, does not change the result of the fair coin. But, somehow, the halfer's retort seems to lose some force if SB is asked not about the coin but about her degree of credence that it is Tuesday. Of course, the halfer answer is 1/4 while the thirders claim 1/3; so what? But when the question is about the awakening, not (directly) about the coin, it seems more natural to count awakenings.

REFLECTION REVISITED. The idea that if you know that tomorrow you will think "X" then you should already think "X" today is known to many philosophers as Bastiaan van Fraassen's reflection principle [13, 14]. The reflection principle is for the most part a special case of the optional stopping theorem (see, e.g., Richard Durrett [11]), which implies that, if whenever a learning process (more generally, a martingale) is stopped the probability of a given event is the same, then the event's *a priori* probability must be that same value. But the stopping algorithm must be implementable. This will not be the case, even when the stopping time is fixed on the clock, if SB doesn't know what time it is. (See, e.g., [38]—another great title.)

In fact, failure of Van Fraassen's reflection principle in such a case is a fairly ordinary occurrence, and no memory loss is required for a demonstration. In Fred's town, if school is to be suspended on account of snow, a loud siren blast is heard at 7:00 a.m. sharp. Fred wakes up much earlier, estimating a probability of 1/2 that a snow day will be declared but has no watch or clock. As time passes and he doesn't hear the blast, his estimate of the snow day probability will go down, reaching perhaps 1/3 (that magic number![7]) at 6:59 a.m. Of course, this number will suddenly jump to 1 at 7:00 a.m. if it is a snow day; otherwise, it will continue decreasing, reaching 0 when Fred is sure it's past 7:00 a.m. But the point is, Fred *knows* that his estimate of the snow day probability will be lower at 6:59 than it is now.

The optional stopping time theorem is not contradicted because, for Fred, "6:59" is not a legitimate stopping time. If instead Fred considers his snow-day-probability-estimate at the point when it becomes light enough to see his dresser—a genuine stopping time—he finds that its expected value is 1/2.

Similarly, "Monday" is not a legitimate stopping time for SB unless she is to be told the day of the week. When she is told that it's Monday, the theorem applies and we correctly conclude that her credence in Heads remains at 1/2.

INFORMATION REVISITED. How does a thirder handle the argument that, upon awakening, SB has no new information to justify changing her Sunday appraisal? Elga himself, as well as Hitchcock, Monton [29], and Vaidman and Saunders [44], concede the lack of information but believe SB's credence in Heads nonetheless changes. Robert Aumann, Sergiu Hart, and Motty Perry [3], giving a thirder argument years before Elga [12] in connection with the Absent-Minded Driver, take this view as well.

Other thirders, especially lately, have argued effectively that to be conscious at a given moment, even if you don't "know what time it is," can constitute genuine information, justifying updating your degree of credence. Included in this list are Arntzenius [2] (after his conversion to thirdism), Dorr [10], Horgan [19], Karlander and Spectre [21], and Weintraub [45].

There are two obvious objections to this contention. One is that, if you weren't conscious, you wouldn't have the information. True, but so what? The Sleeping Beauty problem itself provides an example: If SB wakes up on Tuesday and is told the day, she undoubtedly has information and can use it to conclude Tails. Yet, if she is not awakened on Tuesday, she may never know that the coin came up Heads. Similarly, Alice (in the sleeping pill example) learns upon awakening in the middle of the night that the sleeping pill did not work but would not have received the contrary information.

More persuasive is the idea that for your consciousness to provide information, you must know *when* you are conscious. But this calls into question the arbitrariness of how, as well as whether, time is measured. Must a moment be labeled to be a moment? Suppose that in SB's experimental bedroom is an LED device that tells the number of days since the Chicxulub impact. She awakens and reads the number 24,120,373,498. She has no idea what day of the week that represents, but she does know that she is conscious on that day and didn't know before that she would be.[0]

Seen this way, it would be surprising if SB got no information when awakened. Once it is conceded that being conscious at a time when SB might not have been conscious can justify modifying a credence, it follows that being conscious at what *might* be a time when she might not have been conscious also can justify modifying a credence.

4. Conclusions

Sleeping Beauty is indeed a beauty of a problem, and I am under no illusions that controversy about its solution will ever entirely disappear. But the extensive thought and discussion devoted to Sleeping Beauty has not been in vain; the literature suggests, at least to this writer, the following points.

- The Sleeping Beauty problem touches many fascinating issues in philosophy but, to the extent that there is agreement about what is asked, is also a mathematical question to which many think the straightforward answer is 1/3.
- Some consequences of the 1/3 answer appear surprising at first, but upon scrutiny, seem (for some) increasingly intuitive. In particular, being conscious at a given moment may constitute legitimate information, even if—and in some cases, especially if—the moment's time label is not known.
- Kolmogorov's axioms—in particular, their treatment of conditioning—have held up quite well and together with their implications (e.g., the optional stopping time theorem) may indeed have gained some admirers. Andrey Nikolayevich, were he conscious, would be justly proud. But no comprehensive theory of credence and how it is updated has been agreed upon.

Finally, there is ample confirmation here that philosophers and mathematicians have a lot to gain by talking to one another. The former, for example, are reminded that replacing a principle by a theorem can help achieve clarity, the latter that the path from words to numbers can have hidden twists.

Acknowledgments

The author is greatly indebted to stimulating email conversations with Spencer Backman, Jeff Friedman, Jon Hart, Doug McIlroy, Sam Miner, Jo Ellis-Monaghan, Pradeep Mutalik, Jay Pantone, Tom Zaslavsky, Olga Zhaxybayeva, and especially the very patient Mike Titelbaum. Of course, none of the above can be assumed to endorse this article. Research was supported by NSF grant DMS-0901475.

Notes

1. In its simplest form, the absent-minded driver wants to take the second highway exit to get home but can't distinguish the second exit from the first and knows that he will not remember whether he already passed an exit. Preferring to miss both exits rather than get off at the first, he reluctantly decides to stay on the highway, but when he gets to an exit, he reconsiders. In considering randomized algorithms for the driver, most writers assumed what would be the thirder view—causing, according, e.g., to Wolfgang Schwarz [37], paradoxical conclusions

2. More about causal versus evidential decision theory will be found in the "Gambling revisited" section below.

3. Perspectivalism says that two agents can rationally disagree about a proposition even though each gives the other's argument the same weight as her own. Pittard uses "robust" to emphasize that in his SB variation, the arguments do not merely have equal weight; they are identical.

4. An issue with [6] is that its equation (1) gives the probability that an event is witnessed at least once during the experiment, but what is required for SB's conditioning is that the event is witnessed at time t.

5. which would indeed be the case, were SB not told in advance, in Elga's argument, that she would learn what day it is.

6. It is not my intention to make fun of philosophy here; quite the contrary: I love it that philosophers question everything. Somebody has to!

7. The argument for 1/3 is that at 6:59 Fred's a priori probability that 7:00 a.m. has passed by is (almost) 1/2, now reduced to 1/3 by the condition of the siren not having gone off yet. Thus "before 7:00 a.m., snow day," "before 7:00 a.m., no snow day," and "after 7:00 a.m., no snow day" are equiprobable. Sounds like SB, no? But (as Arntzenius points out in [2]) Fred can't know what his a priori time probability distribution will be at a given time; otherwise, he could use that information to tell time!

8. At this point in the article, readers should not be surprised to learn there are frameworks—see, e.g., [6, 16, 28]—in which the presence of any time-measuring device changes the answer to the Sleeping Beauty problem. Others, e.g., Kierland and Monton [22], say that unknown time is not really the issue to begin with; the Monday/Tuesday factor can be replaced by a random signal, such as the color of the room in which SB is awakened.

References

1. F. Arntzenius, Reflections on Sleeping Beauty, *Analysis* **62** no. 1 (2002) 53–62.

2. ———, Some problems for conditionalization and reflection, *J. Philosophy* **199** no. 7 (2003) 356–370.

3. R. J. Aumann, S. Hart, M. Perry, The forgetful passenger, *Games Econom. Behav.* **20** (1997) 117–120.

4. N. Bostrom, Sleeping Beauty and self-location: A hybrid model, *Synthese* **157** no. 1 (2007) 59–78.

5. R. Briggs, Putting a value on Beauty, *Oxford Studies in Epistemology* **3** (2010) 3–34.

6. J. Cisewski, J. B. Kadane, M. J. Schervish, T. Seidenfeld, R. Stern, Sleeping Beauty's credences, *Philos. Sci.* **83** no. 3 (2016) 324–347.

7. V. Conitzer, A Dutch book against Sleeping Beauties who are evidential decision theorists, *Synthese* **192** no. 9 (2015) 2887–2899.

8. ———, Can rational choice guide us to correct *de se* beliefs? *Synthese* **192** no. 9 (2015) 4107–4119.

9. M. Cozic, Imaging and Sleeping Beauty: A case for double-halfers, *Internat. J. Approx. Reason.* **52** (2011) 137–143.

10. C. Dorr, Sleeping Beauty: In defence of Elga, *Analysis* **62** (2002) 292–296.

11. R. Durrett, *Probability: Theory and Examples*. Wadsworth, Pacific Grove, CA, 1991.

12. A. Elga, Self-locating belief and the Sleeping Beauty problem, *Analysis* **60** no. 2 (2000) 143–147.

13. B. C. van Fraassen, Belief and the will, *J. Philosophy* **81** (1984) 235–256.

14. ——, Belief and the problem of Ulysses and the Sirens, *Philos. Stud.* **77** (1995) 7–37.

15. B. Groisman, The end of Sleeping Beauty's nightmare, *British J. Philos. Sci.* **59** (2008) 409–416.

16. J. Y. Halpern, Sleeping Beauty reconsidered: Conditioning and reflection in asynchronous systems, *Oxford Studies in Epistemology* **3** (2005) 111–142.

17. P. Hawley, Inertia, optimism and Beauty, *Noûs* **1** no. 1 (2013) 85–103.

18. C. Hitchcock, Beauty and the bets, *Synthese* **139** no. 3 (2004) 405–420.

19. T. Horgan, Sleeping Beauty awakened: New odds at the dawn of the new day, *Analysis* **64** (2004) 10–21.

20. ——, Synchronic Bayesian updating and the generalized Sleeping Beauty problem, *Analysis* **67** no. 1 (Jan. 2007) 50–59.

21. K. Karlander, L. Spectre, Sleeping Beauty meets Monday, *Synthese* **174** (2010) 397–412.

22. B. Kierland, B. Monton, Minimizing inaccuracy for self-locating beliefs, *Philosophy and Phenomenological Research* **70** (2005) 384–395.

23. N. Kim, Sleeping Beauty and shifted Jeffrey conditionalization, *Synthese* **168** (2009) 295–312.

24. A. N. Kolmogorov, *Foundations of the Theory of Probability*. Ed. N. Morrison. Chelsea, New York, 1950.

25. D. Lewis, Sleeping Beauty: Reply to Elga, *Analysis* **61** (2001) 171–176.

26. P. J. Lewis, Quantum Sleeping Beauty, *Analysis* **67** no. 293, 59–65.

27. I. Mariolis, Revealing the beauty behind the Sleeping Beauty problem (2014), http://arxiv.org/pdf/1409.3803.pdf.

28. C. J. G. Meacham, Sleeping Beauty and the dynamics of *de se* beliefs, *Philos. Stud.* **138** no. 2 (2008) 245–269.

29. B. Monton, Sleeping Beauty and the forgetful Bayesian, *Analysis* **62** no. 1 (Jan. 2002) 47–53.

30. P. Mutalik, Why Sleeping Beauty is lost in time, *QUANTA* (March 31, 2016), http://quantamagazine.org/20160331-why-sleeping-beauty-is-lost-in-time.

31. M. Piccione, A. Rubinstein, On the interpretation of decision problems with imperfect recall, *Games Econom. Behav.* **20** (1997) 3–24.

32. ——, The absent-minded driver's paradox: Synthesis and responses, *Games Econom. Behav.* **20** (1997) 121–130.

33. J. Pittard, When Beauties disagree: Why halfers should affirm robust perspectivalism. *Oxford Studies in Epistemology* **5** (Feb. 2015).

34. J. Pust, Horgan on Sleeping Beauty, *Synthese* **160** (2008) 97–101.

35. J. S. Rosenthal, A mathematical analysis of the Sleeping Beauty problem, *Math. Intelligencer* **31** (2009) 32–37.

36. J. Ross, Sleeping Beauty, countable additivity, and rational dilemmas, *Philos. Rev.* **119** (2010) 411–447.

37. W. Schwarz, The Absentminded Driver: No paradox for halfers, (draft, February 11, 2008), http://www.umsu.de/words/driver.pdf.

38. T. Seidenfeld, M. J. Schervish, J. B. Kadane, Stopping to reflect, *J. Philosophy* **101** no. 6 (2004) 315–322.

39. S. Tadelis, *Game Theory*, Princeton Univ. Press, Princeton, NJ, 2013.

40. M. Teitelbaum, *Sleeping Beauty* (Disney Princess ed.). Little Golden Books, New York, 2004.

41. M. G. Titelbaum, The relevance of self-locating beliefs, *Philos. Rev.* **177**, 555–605.

42. ——, Ten reasons to care about the Sleeping Beauty problem, *Philos. Compass* **8** no. 11 (2013) 1003–1017.

43. ——, Self-locating credences, in *The Oxford Handbook of Probability and Philosophy*. Ed. A. Hajek, C. Hitchcock. Oxford Univ. Press (2016).

44. L. Vaidman, S. Saunders, On Sleeping Beauty controversy (2009), http://philsciarchive .pitt.edu/archive/00000324.

45. R. Weintraub, Sleeping Beauty: A simple solution, *Analysis* **64** (2004) 8–10.

46. R. White, The generalized Sleeping Beauty problem: A challenge for thirders, *Analysis* **66** (2006) 114–119.

47. A. Wilson, Everettian confirmation and Sleeping Beauty, *British J. Philos. Sci.* **65** no. 3 (1 Sept. 2014).

Wigner's "Unreasonable Effectiveness" in Context

José Ferreirós

Einstein famously wrote that the most incomprehensible thing about the world is that it is comprehensible. He was thinking about mathematical and theoretical physics. The idea is an old one. Nobel Prize winner Paul Dirac believed that mathematics was an especially well-adapted tool to formulate abstract concepts of any kind, and he also famously insisted that mathematical beauty is a key criterion for physical laws.[1] But one of the most famous presentations of that thought was by Dirac's brother-in-law, Wigner Jenő Pál, a.k.a. Eugene P. Wigner.

Wigner was a highly successful scientist. In mathematical circles, he is best known for his contributions to quantum theory, pioneering the application of group theory to the discovery of fundamental symmetry principles—and, of course, for his 1960 paper "The Unreasonable Effectiveness of Mathematics in the Natural Sciences." Some passages of the 1960 paper are often quoted; here is one:

> The miracle of the appropriateness of the language of mathematics for the formulation of the laws of physics is a wonderful gift which we neither understand nor deserve. We should be grateful for it and hope that it will remain valid in future research and that it will extend, for better or for worse, to our pleasure even though perhaps also to our bafflement, to wide branches of learning (Wigner 1960, 237/549).

Toward the beginning of his essay, Wigner writes that "the enormous usefulness of mathematics in the natural sciences is something bordering on the mysterious and [. . .] there is no rational explanation for it" (1960, 223/535). It is telling that the word "miracle" appears twelve times in the text!

Surely Wigner's focus was more on the question of what it is *in the physicist's approach* to reality, as it has developed since Newton, that makes it possible to formulate mathematical laws.[2] But in fact his paper has been widely discussed in connection with a related question, which is our concern here; namely, what is it *within mathematics* that makes possible its highly successful application in physics? A good number of people have offered replies to Wigner, aiming to show that there is no miracle.[3] Here too we shall critically discuss elements of Wigner's presentation that unduly transform the relation between mathematics and physics into "a gift" or "miracle" that is very difficult to understand. Beyond that, we shall try to unveil the sources of Wigner's point of view. His discussion is characteristic of mid-twentieth-century images of mathematics, but it is hard to square with present conceptions or even with the views of the best experts from a generation before him.[4]

Wigner's Views and the New Practice of Mathematical Physics

I will distinguish three different parts in Wigner's 1960 paper. First, there are three sections devoted to generalities about mathematics and physics, in which the reflections regarding physics stand out as more relevant and insightful. Concerning physics, he lays emphasis on how the identification of regularities in the chaotic phenomena depends on packing a lot of the information into the "initial conditions."[5] Next is a section in which Wigner makes his strongest case, highlighting the success of mathematical laws in physical theories to underscore how it is "truly surprising." Finally, he moves on to question the uniqueness of physical theory, that is, the hope for a single foundation of all physics or even all science.[6] I will begin in the middle by explaining the strongest case Wigner makes for the astonishing effectiveness of mathematics as a central component of the methodology of physics.

In the section entitled "Is the Success of Physical Theories Truly Surprising?" Wigner offers three examples—which, he adds, could be multiplied almost indefinitely—to illustrate the appropriateness and accuracy of the mathematical formulation of the laws of nature:

- Newton's law of gravitation,
- Heisenberg's rules of matrix mechanics, and
- the theory of Lamb shift in QED.

The law of gravity, "which Newton reluctantly established" and which he could verify to within an error of about 4 per cent, has proved to be accurate to within an error of less than 1/10,000 of 1 per cent. It "became so closely associated with the idea of absolute accuracy that only recently did physicists become again bold enough to inquire into the limitations of its accuracy" (Wigner 1960, 231/543).

As for the second example, from the early years of quantum mechanics, Heisenberg established some quantum-mechanical rules of computation—which were to lead to matrix mechanics—on the basis of a pool of data that included the behavior of the hydrogen atom and its spectrum. When Pauli (1926) applied quantum mechanics to the hydrogen atom in a realistic way, the positive results were an expected success. But then, says Wigner, it was "applied to problems for which Heisenberg's calculating rules were meaningless." These rules presupposed that the classical equations of motion had solutions with certain periodicity properties, "and the equations of motion of the two electrons of the helium atom, or of the even greater number of electrons of heavier atoms, simply do not have these properties, so that Heisenberg's rules cannot be applied to these cases. Yet the calculation of the lowest energy level of helium, as carried out a few months ago [in 1959] by Kinoshita at Cornell and by Bazley at the Bureau of Standards, agree with the experimental data within the accuracy of the observations, which is one part in ten millions. Surely in this case we got something out of the equations that we did not put in" (Wigner 1960, 232/544).

Certainly Wigner has a point here. In his view, it is the theoretical separation between initial conditions of the system and the simple, mathematical "laws of nature" that have allowed physicists to attain such impressive levels of success. Wigner is right in that the actual empirical success of physical laws went far beyond anything that might reasonably have been expected at the outset. It is unconvincing to regard this as the outcome of mere chance, but is there "no rational explanation for it"?

As we shall see, the way in which Wigner framed his understanding of mathematics plays a large role in creating the "mystery," the impression of a miracle. The advancement of physical science shows undeniably that there are mathematical structures underlying natural processes and phenomena. (Of course we lack an *a priori* argument that it *must* be so, but science never offers ultimate answers.) Even if we admit that there is a common structure between our mathematical models and real phenomena, this does not force us to interpret realistically *all features* of the models. That is, one can still be critical and ponder the possibility that some features of the mathematics may be human artifacts that perhaps impute extra structure, complications which distance our physico-mathematical understanding from "the real" itself.

Wigner was an important figure in the emergence of the radically new mathematical toolbox of quantum physics, built on top of new, abstract, unintuitive representations. Some physicists resented the abandonment of the tool kit of classical analysis in favor of group-theoretic methods, abstract spaces, and so on. Around 1930, they described these innovations as a "group pest" or "plague of groups." The situation worsened when, instead of seeking explicit solutions by calculus, the new goal became to find invariants associated with structural representations. Higher levels of theorizing began to occupy center stage; a case in point was symmetry considerations one level above the mathematical laws of physics (Scholz 2006).

Broadly speaking, an essential ingredient of the new type of work was the infusion of a new style of structural and qualitative methods (set theory, topology, symmetries) to replace the old quantitative spirit and its search for concrete solutions on the side of calculation. Little wonder that questions would arise about the new balance between mathematics and physics. Before we discuss his views on mathematics, I will argue that Wigner's formative years in Berlin seem to have been particularly relevant in shaping his philosophical views.

Some Biographical Elements

Eugene Wigner, who was born in Budapest in 1902, came from a well-to-do family. His long life included an extended period in Berlin until 1936, and a still longer one in the United States, mostly working at Princeton. He was awarded the Nobel Prize for Physics in 1963 "for his contributions to the theory of the atomic nucleus and the elementary particles, particularly through the discovery and application of fundamental symmetry principles."[7] Obviously the Nobel increased the visibility of Wigner's reflections on science, and in 1967 he published a selection of essays under the apt title *Symmetries and Reflections*. One of them was "The Unreasonable Effectiveness of Mathematics in the Natural Sciences," originally published in the *Annals of Pure and Applied Mathematics* in 1960.

In Budapest, Wigner attended a secondary school (*Gimnázium*) where he obtained a sound training in mathematics. Two people were crucial in this respect, the noted mathematics teacher László Rátz, who knew how to care about promising students, and a fellow student who was one year younger, Neumann János, a.k.a. John von Neumann. Wigner's friendship with von Neumann was a lasting one, and he would later acknowledge that he learned more mathematics from von Neumann than from anyone else.[8]

Wigner studied chemical engineering at the Technical University in Berlin, a choice strongly influenced by his father. He himself was more attracted to physics, and this led him to attend the Wednesday meetings of the German Physical Society, where he could see and hear luminaries such as Einstein, Planck, Sommerfeld, and Heisenberg. A noteworthy remark in his autobiography reads, "In my apartment, I read books and articles on chemical analysis, set theory, and theoretical physics" (Szanton 1992, 65). His independent reading on set theory is noteworthy, but this was probably because von Neumann was heavily engaged with the subject.

In the academic year 1926–1927, Wigner obtained a position in Berlin as an assistant to Karl Weissenberg, who worked on X-ray crystallography. Through his engagement with crystallography, Wigner was led to study group theory, taking up the algebra textbook by Heinrich Weber and then solving questions posed by Weissenberg.[9] The following academic year, he went to Göttingen to work as an assistant to David Hilbert. This might have been a momentous opportunity, yet things did not work out so well since Hilbert was seriously ill. Wigner was left to work on his own, and so he decided to investigate the relation between group theory and the new quantum mechanics. Von Neumann had given him a crucial pointer, suggesting that he use group representations as found in the relevant papers by Georg Frobenius and Issai Schur.[10] Thus he became a pioneer in the new mathematical methods of theoretical physics.

Someone (presumably Wolfgang Pauli) characterized the period we are talking about as a time of "Gruppenpest" in physics.[11] The metaphor of a disease reflects the feeling of alienation experienced by many theoretical physicists, realizing that their traditional toolbox of classical analytical methods was being replaced by new and foreign "abstract" ideas. Wigner worked especially on the study of atomic spectra, which was to be the topic of his important book *Gruppentheorie und ihre Anwendungen auf die Quantenmechanik der Atomspektren* (1931). In the introduction, he emphasizes how the precise solution of quantum mechanical equations by calculus is extraordinarily difficult, so that one could only obtain gross approximations. "It is gratifying, therefore, that a large part of the relevant results can be deduced by considering the fundamental symmetry operations [*durch reine Symmetrieüberlegungen*]." He adds,

> Against the group-theoretic treatment of the Schrödinger equation, one has often raised the objection that it is "not physical." But it seems to me that a conscious exploitation of elementary symmetry properties ought to correspond better to physical sense than a treatment by calculation.[12]

In 1928, Wigner became a *Privatdozent* at the Technische Hochschule in Berlin, but given the worsening political situation in Europe, in 1936 Wigner and von Neumann decided to settle permanently in "the New World." Nevertheless, the European years in the 1920s and 1930s had a particularly strong impact on Wigner's views. In Germany at the time, there was an intense sense of rupture, of new forms of life being created. Quantum mechanics was perceived as a radical break with the past. One spoke of "Knabenphysik," because its protagonists were all "youngsters" (except for Max Born and Niels Bohr). This tense social and intellectual atmosphere was alluded to in Wigner's reminiscences:

> Historians tell us that Berlin in the 1920s was a city in chaos. . . . Is a radical a man who repudiates the society of his parents and teachers? If so, then I was no radical in Berlin. I admired my teachers more with each passing year. I loved my parents and wanted to help them. To dream of pursuing a career that they had not chosen was a radical enough path for a youth of my background. I had no wish to be more radical than that. But if a radical is someone who regards a traditional subject in a revolutionary way, then perhaps I was a radical, because quantum mechanics had transformed physics and I embraced quantum mechanics fervently (Szanton 1992, 84).

Wigner's early exposure to abstract mathematical theories led him to adopt some new and radical ideas about mathematics. These ideas were very different from the views of previous generations, and they came to be clearly expressed in his 1960 paper.

What Is Mathematics? To Be or Not To Be a Formalist

In the section "What is mathematics?" Wigner provides a surprisingly simple answer to this question:

> [. . .] mathematics is the science of skillful operations with concepts and rules invented just for this purpose. The principal emphasis is on the invention of concepts. [. . .] The depth of thought which goes into the formulation of the mathematical concepts is later justified by the skill with which these concepts are used (Wigner 1960, 224/536).

Wigner here emphasizes the predominant role of intramathematical considerations, regardless of the potential for application to real phenomena. However, he makes a distinction. On the one hand, we have basic ideas such as the concepts and principles of elementary geometry, rational arithmetic, and even irrational numbers—which are directly suggested by the physical world. On the other hand,

> Most more advanced mathematical concepts, such as complex numbers, algebras, linear operators, Borel sets—and this list could be continued almost indefinitely—were so devised that they are apt subjects on which the mathematician can demonstrate his ingenuity and sense of formal beauty (Wigner 1960, 224/536).

Ingenuity, inventiveness, the skill of the virtuoso to develop interesting connections, guided by a sense of formal beauty and a basic concern for logical coherence, are what drive pure mathematics.

Such a description brings to mind the modernist work of von Neumann around 1930: developing axiomatic set theory in a way completely different from Zermelo's and introducing the notion of Hilbert space to redefine in a much more abstract setting the foundations of quantum mechanics.[13] Further, von Neumann was working on Hilbert's metamathematical program and was invited to represent the foundational standpoint of formalism in a conference on the Epistemology of the Exact Sciences, organized jointly by the Berlin Society for Empirical Philosophy and the Vienna Circle in September 1930.

Strict formalism interprets mathematical systems as a game of symbols. The symbols have no other content than they are assigned in the calculus by their behavior with respect to certain rules of combination; the only requirement is consistency of the system. The network of relations thus codified in a formal calculus restricts possible applications or *interpretations* of the system. In the 1930s, this formalistic standpoint had the advantage of eliminating all "metaphysical difficulties" concerning mathematics, and in particular the need for positing any Platonic realm of mathematical objects. Formalism also incorporated traits of "conventionalism" about mathematics, such as the insistence on simplicity and elegance or beauty as guides in the formulation of the basic principles of axiomatic systems.

In fact, Wigner's paper offers very little information on his sources about the idea of mathematics. We find a brief reference to Hilbert on foundations (1922a), another passing reference to Karl Polanyi,[14] and mention of *Die Philosophie der Mathematik in der Gegenwart* (1932) by Walter Dubislav. Yet if we reflect on these sources, adding Wigner's time in Göttingen and his relation to von Neumann, links to the Hilbert School are predominant.

The case of Walter Dubislav (1895–1937) is particularly interesting. He was a member of the Berlin Association for Empirical Philosophy and one of the signers of the famous Vienna Circle manifesto. He began studying mathematics at Göttingen, but World War I intervened. After military service, he went to the University of Berlin, concentrating on philosophy and logic. The brief presentation of the philosophy of mathematics offered in his 1932 textbook is very clear, emphasizing mathematical logic, axiomatic thinking, and a form of empiricism in the case of applied mathematics. The imprint of the Hilbert School is undeniable here. Dubislav argued for the "character of calculation [*Kalkülcharakter*] in pure mathematics" and defended a strict formalism:

Formalism states the following: that pure logic like pure mathematics are in the strict sense of the term not sciences, [. . .] Pure logic and pure mathematics are calculi [*Kalküle*] which deal with this: obtaining from certain initial formulas, arbitrary in themselves, more and more formulas according to rules of operation that in themselves are arbitrary. Put grossly: pure logic and pure mathematics, taken in themselves, are games of formulas [*Formel-spiele*] and nothing else (Dubislav 1932).

This was not yet a familiar point of view. As it turns out, Dubislav was a *Privatdozent* at the Technische Hochschule Berlin from 1928 and a colleague of Wigner there.[15]

Two issues deserve to be emphasized. First, logical empiricism would continue to be prominent in the philosophical context around Wigner and is visible in his (1960). But the second point is more directly interesting for my purposes. We have seen that Dubislav was a strict formalist, and that Wigner himself still defended a kind of formalism in his remarks about mathematics. The previous generation of physicists and mathematicians were not formalists, and it was only the generation that matured in the 1920s that understood the new ideas about axiomatics, structures, logic, and foundations in a radical way. The situation is parallel to the radicalism of the new conceptions of the physical world among the "youngsters" who advanced quantum physics.

Hilbert was not a formalist at the level of epistemology. His celebrated formalism was a *method* adopted in the context of studies of the foundations of mathematics, for the goals of metamathematics (consistency proofs, decision procedures). Using the axiomatic method, one may begin by considering a particular field of work with concrete ideas. But there is much to gain methodologically by disregarding the particular meaning of the concepts, considering the axioms as schematic conditions, and adopting full freedom of interpretation. In foundational research, this attitude can be amplified to achieve strict formalization, but these methods do not expand into a full epistemological account, and such an account was not at all Hilbert's intention.

Incidentally, it is easy to find a thousand places in which Hilbert is alleged as saying, "Mathematics is a game played according to certain simple rules with meaningless marks on paper." The source of this (mis)quotation seems to be E. T. Bell, and it can nowhere be found in Hilbert's papers. What we can find in lectures of 1919–1920 is the following: "There is no talk of arbitrariness here. Mathematics is not like a game in which the problems are determined by rules invented arbitrarily, but a conceptual system [endowed] with inner necessity, that can only be this, and not any other way."[16]

I have previously mentioned the novelty of the work in mathematical physics around 1930 with its infusion of a new spirit of structural and qualitative methods in place of the old quantitative spirit. But this was not unknown to Henri Poincaré or Hilbert, so it cannot be simply regarded as the source of a formalist attitude. In this case, other more general sources have to be found, coming largely from the intellectual context.

As we discussed earlier, the young intellectuals in Berlin, during the 1920s, were living in a rather chaotic, rapidly changing, and heated cultural atmosphere. After wartime defeat and the political and economic turmoil (inflation, the Weimar republic), one could hear everywhere the call for a "new order," a new society, indeed a "new man," and of course new forms of science. Such a setting promoted forms of *modernism* in the sciences, modernistic tendencies that presented themselves as a radical break with the past.[17] Jeremy Gray (2008) offered a reconstruction of early twentieth-century mathematics as undergoing a "modernist transformation." He defines modernism, in science or mathematics, as the new conception of the field as "an autonomous body of ideas, having little or no outward reference, placing considerable emphasis on formal aspects of the work and maintaining a complicated—indeed, anxious—rather than a naïve relationship with the day-to-day world" (Gray 2008, 1). The interwar period was a particularly high time for such modernist tendencies.

Hilbert was basically right: formalism is very good as a method for studying foundations, but philosophical questions about the epistemic nature of mathematical knowledge require more sophisticated answers. Moreover, the examples Wigner presents from advanced mathematics do not support his formalist views. His list included complex numbers, algebras, linear operators, and Borel sets. His idea was that the elaboration of such concepts is guided by intramathematical considerations, disregarding considerations of the potential for application to natural phenomena. They are, according to him, "so devised that they are apt subjects on which the mathematician can demonstrate his ingenuity and sense of formal beauty."

What Is Mathematics? Remarks on the Reasonable Development from Physics

Wigner's most convincing example is that of complex numbers. Italian mathematicians introduced the square root of -1 in the sixteenth century to manipulate numbers and expressions in algebraic equations. Astonishingly, the imaginary numbers turned out to play important roles in different places: the equation that some consider the most beautiful in all of math, $e^{\pi i} + 1 = 0$ (Euler), the fundamental theorem of algebra (D'Alembert and Gauss), Cauchy's integral formula, Riemann's mapping theorem, etc. Even so, the early results and procedures did not establish a secure position for imaginary numbers in the world of mathematics. Their full adoption occurred only in the nineteenth century and involved a reconception of the number concept as well as the establishment of geometric representations of the complex numbers. For Gauss and Riemann, considering the system of complex numbers as the *natural* general framework for number in general was a basic commitment and fundamental principle of pure mathematics.[18]

Wigner later emphasizes that quantum mechanics is formulated on the basis of *complex* Hilbert space. This is why theoretical physicists, such as Wigner and Roger Penrose, have placed great emphasis on the complex number structure. However, going somewhat against Wigner's thesis, the complex field inherits most of its properties from the real field. Wigner himself stressed that the real number system was devised so as to mirror the properties of measurable quantities. But let us concede that the story of complex numbers fits well with Wigner's viewpoint; their importance in connection with the theory of electromagnetic fields and, even more, quantum theory is astonishing.

Given that Wigner pioneered group-theoretic methods in quantum mechanics, it is noteworthy that group theory is not among his examples. However, it may well be the case that Wigner was aware of opinions like Hermann Weyl's (1928), that the group concept is in a sense "one of the oldest" mathematical concepts. The reasoning behind this statement is that group structures are implicit behind all kinds of ancient concepts and practices—symmetry considerations, operations of translation and congruence in basic geometry, measuring operations, and so on. Weyl's quite reasonable view is that nineteenth-century explorations and formalizations just made explicit and abstract what had been there, implicitly, throughout the history of mathematics.

Likewise, linear operators and matrices might have seemed a very novel feature to physicists around 1930, for the simple reason that they had not been part of their basic education, but in fact linear algebra arose naturally in different areas of mathematics *and* its applications. As Kleiner (2007, 79) remarks, "the subject had its roots in such diverse fields as number theory (both elementary and algebraic), geometry,[19] abstract algebra (groups, rings, fields, Galois theory), analysis (differential equations, integral equations, and functional analysis), and physics." Thus these examples were not good choices for Wigner's purposes.

Perhaps the oddest example in Wigner's list is his reference to Borel sets, given that these play no immediate role in physics. Probably Wigner chose the example of Borel sets as one of the central concepts of set theory in the first third of the twentieth century—what better example of the purest in pure math?[20] Yet this case goes rather against his thesis. Borel sets are strongly linked with the function concept, and their study was motivated by a desire to *restrain* the most general and arbitrary possibilities opened by set theory, to focus on concrete ideas closer to classical math.[21] The all-important notion of function is something that one misses in Wigner's list. But the study of functions has constantly been promoted by *extra*mathematical considerations, mostly physical.

As suggested previously, it is natural to compare Wigner's views with Poincaré's. Both were pioneers in the new mathematical methods and their use in physics—the group concept was a key guiding element—and both were highly influential in promoting new qualitative approaches and techniques. Also, both scientists were inclined to general philosophical reflection, and it is interesting that both emphasized the importance of the aesthetic element in guiding pure mathematics. Yet Poincaré never suggests a "miracle" in the role of mathematics in physics; on the contrary, he insisted on the interplay between mathematics and science, and he (unlike Wigner) emphasized the centrality of the continuum and the function concept. Thus in *The Value of Science* (1905) he writes,

> physics has not only forced us to choose among problems which came in a crowd; it has imposed upon us problems such as we should without it never have dreamed of.[22]

A case in point might be Fourier's work in *Théorie analytique de la chaleur* (1822), in which he used trigonometric series in mathematical physics, also linked with the famous eighteenth-century discussion about vibrating strings. Fourier series were the background for Dirichlet's (1829) proposal of the notion of arbitrary function, as well as Riemann's (1854) study of highly discontinuous functions and his notion of the integral.

Conclusions

Even Wigner's friend, János von Neumann, who may have entertained modernist views akin to formalism around 1930, was no longer in agreement with him after World War II. In an interesting paper for the general public, published in 1947, he writes,

> The most vitally characteristic fact about mathematics is, in my opinion, its quite peculiar relationship to the natural sciences, or, more generally, to any science which interprets experience on a higher than purely descriptive level. [. . .] Some of the best inspirations of modern mathematics (I believe, the best ones) clearly originated in the natural sciences. The methods of mathematics pervade and dominate the "theoretical" divisions of the natural sciences.

Contemporary Soviet mathematicians, who would have regarded Wigner's presentation as a quintessential example of bourgeois philosophy, were even more in favor of such views. I am led to mention this because Wigner, like his Hungarian friends Leo Szilard, Edward Teller, and von Neumann, was strongly anticommunist—and it may be the case that his political views colored his philosophical ideas.

Many of the great abstractions introduced in mathematics from the mid-nineteenth century have strong roots in the (physically motivated) mathematics of functions, analysis, the real-number continuum, and geometry. Actually the twentieth-century abstractions are often based on making the basic assumptions behind the earlier systems *more flexible*. And this increase in flexibility provides a very rational explanation of the applicability of mathematics! It is hardly surprising that a much more general and flexible theory of geometrical structures (e.g., Riemannian differential geometry) can be applied in many contexts in which the rigid structures of Euclidean geometry would not be applicable.

Yet perhaps the empirical success of mathematical laws in physics requires something else.

Notes

1. Dirac (1963) expressed this consideration in very strong terms; see also Kragh (1990).
2. In an interesting paper, Arezoo Islami (2016) suggests that this is the right understanding of Wigner's paper. This convincing reading would also help to explain the relatively

careless presentation of ideas about mathematical theory, as opposed to physical theory, in the paper

3. An example in Mathematical Intelligencer is Grattan-Guinness 2008; see also Lützen (2011) and Russ (2011).

4. For a broad and enlightening historical perspective on this topic, see Bottazzini and Dalmedico (2001).

5. Wigner's views on physical theory are very interesting, but we cannot go into details here. The interested reader may consult his Nobel lecture, in which he amplifies these themes, and also Islami (2016).

6. He insists that it is conceivable that one will be unable to unify the fundamental physical theories, and even more so for theories of biology or of consciousness. This argument may well have been aimed at the Unity of Science movement, which was seeking to unify all science from a physicalist standpoint.

7. Half the prize went to Wigner, and the other half jointly to Maria Goeppert Mayer (the second woman to get the prize, after Marie Curie) and to J. Hans D. Jensen.

8. "Jancsi von Neumann taught me more mathematics than any other of my teachers, even Ratz of the Lutheran gimnázium. And von Neumann taught not only theorems, but the essence of creative mathematical thought: methods of work, tools of argument" (Szanton 1992, 130).

9. See Szanton (1992, 105–106); one of these questions was recognized by von Neumann to be related to group representations, and he told Wigner to study Frobenius and Schur. On Weber's textbook, a crucial source for one or two generations of algebraists, see Corry (1996).

10. Such as Schur (1905). See Hawkins (2000).

11. Szanton (1992, 116–117). In a 1963 interview with Kuhn, Wigner said, "I don't think [Pauli] liked it particularly . . . there was a word, Die Gruppenpest, and you have to chase away the Gruppenpest. But Johnny Neumann told me, 'Oh these are old fogeys; in five years every student will learn group theory as a matter of course,' and essentially he was right." (Kuhn 1963)

12. Years later, when the English version was published, he wrote, "When the original German version was first published, in 1931, there was a great reluctance among physicists toward accepting group theoretical arguments and the group theoretical point of view. It pleases the author that this reluctance has virtually vanished in the meantime and that, in fact, the younger generation does not understand the causes and the basis for this reluctance. Of the older generation it was probably M. von Laue who first recognized the significance of group theory as the natural tool with which to obtain a first orientation in problems of quantum mechanics."

13. According to Saunders Mac Lane (in Duren 1989, 330), after a lecture by von Neumann at Göttingen in 1929, Hilbert asked "Dr. von Neumann, ich möchte gern wissen, was ist dann eigentlich ein Hilbertscher Raum?" ("Dr. von Neumann, I would like to know, what after all is a Hilbert space?").

14. There is not enough space here to develop, but one should emphasize that Polanyi was a very important influence, "my dearest teacher" who "decisively marked my life" (Szanton 1992, 76). A physical chemist, Polanyi was to become a philosopher of science and may have influenced Wigner insofar as he was heavily marked by matter/mind dualism (Szanton 1992, 76 ff). See also Esfeld (1999).

15. He may have given the 1930 book to Wigner as a gift. Wigner (1960, 237/549) also refers to Dubislav's *Natural Philosophy* of 1933, a text defending an empiricist philosophy of science.

16.D. Hilbert (1922b), lectures delivered in 1919–1920. The point has been made repeatedly by experts such as Corry (1996), Rowe (2000), and Mancosu (2010, 139–140).

17. See Gray (2008), Epple and Müller (2019), and earlier work by Herbert Mehrtens (1990).

18. On the history of complex numbers, see Nahin (1998), Ebbinghaus et al. (1991), and Flament (2003).

19. Thus Grassmann in 1844 coming from geometry, and Dedekind in 1871 from algebraic number theory, were among the first to articulate modern ideas about the subject clearly (Kleiner 2007, 84–88).

20. If Wigner studied Hausdorff's (1914) textbook in the 1920s, he must have learned about Borel sets. Hausdorff and Alexandroff proved in 1916 that the continuum hypothesis is true in the limited case of Borel sets; certain properties called "regularity properties" were established for them (e.g., Lebesgue measurability), and set-theorists were hard at work studying how far those properties applied.

21. As a matter of historical fact, Émile Borel, René Baire, and Henri Lebesgue were all critics of Zermelo set theory. After 1905, they all criticized the most general notions of "arbitrary" set, "arbitrary" function, and the axiom of choice. It was their intention to obtain more clarity about the notion of set by focusing on sets that can be "constructed" by well-understood operations. See Ferreirós (1999, 315–316) and the letters from 1905 that were translated in the Appendix of the book by Moore (1982).

22. "Analysis and Physics," Chapter V of Poincaré (1905, 80).

References

Umberto Bottazzini, Amy Dahan Dalmedico (2001). *Changing Images in Mathematics: From the French Revolution to the New Millennium.* London, New York: Routledge.

Leo Corry (1996). *Modern Algebra and the Rise of Mathematical Structures.* Basel, Switzerland; Boston; Berlin: Birkhäuser. 2nd ed. 2004.

Richard Dedekind (1871). Über die Komposition der binären quadratischen Formen, Supplement X to P. G. Dirichlet, *Vorlesungen über Zahlentheorie*, 2nd ed. R. Dedekind, Braunschweig, Vieweg. An English version can be found online at https://www.andrew.cmu.edu/user/avigad/Papers/ideals71.pdf.

Peter Gustav L. Dirichlet (1829). Sur la convergence des séries trigonométriques qui servent à représenter une fonction arbitraire entre des limites données, *J. reine und ang. Mathematik* **4**, 157–169. In *Werke*, Berlin, Reimer, 1889, vol. 1, 283–306.

Paul Dirac (1963). The Evolution of the Physicist's Picture of Nature, May 1963 issue of *Scientific American.*

Walter Dubislav (1932). *Die Philosophie der Mathematik in der Gegenwart.* Berlin: Junker & Dünnhaupt.

—— (1933). *Naturphilosophie.* Junker & Dünnhaupt, Berlin.

Peter Duren, ed. (1989). *A Century of Mathematics in America*, Part 1. Providence, RI: AMS.

Heinz-Dieter Ebbinghaus, Hans Hermes, Reinhold Remmert (1991). *Numbers.* New York: Springer.

Moritz Epple, Falk Müller (2019). *Science as Cultural Practice, Vol. II: Modernism in the Sciences, ca. 1900–1940.* Munich, Germany: De Gruyter.

Michael Esfeld (1999). Essay Review: Wigner's View of Physical Reality, *Studies in History and Philosophy of Modern Physics* 30B, pp. 145–154.

José Ferreirós (1999). *Labyrinth of Thought: A History of Set Theory*. Basel, Switzerland: Birkhäuser.

Dominique Flament (2003), *Histoire des Nombres Complexes*, Paris: CNRS Éditions.

Joseph Fourier (1822). *Théorie analytique de la chaleur*. Paris, Firmin Didot.

Hermann Grassmann (1844). *Die Lineale Ausdehnungslehre*, Leipzig, Teubner. Also in Grassmann, *Gesammelte mathematische und physikalische Schriften*, vol. 1 (1894), New York, Chelsea, 1969, 1–319.

Ivor Grattan-Guinness (2008). Solving Wigner's Mystery: The Reasonable (Though Perhaps Limited) Effectiveness of Mathematics in the Natural Sciences, *Mathematical Intelligencer* 30:3, 7–17.

Jeremy Gray (2008). *Plato's Ghost. The Modernist Transformation of Mathematics*. Princeton, NJ: Princeton Univ. Press.

Felix Hausdorff (1914). *Grundzüge der Mengenlehre*, Leipzig, Germany: Veit. Reprinted by Chelsea in 1949.

Thomas Hawkins (2000). *Emergence of the Theory of Lie Groups: An Essay in the History of Mathematics, 1869–1926*. New York: Springer.

David Hilbert (1922a). Die logischen Grundlagen der Mathematik, *Math. Annalen* 95: 161–165. In Hilbert's *Abhandlungen*, vol. 3, 1935, 178–191.

David Hilbert (1922b). *Natur und mathematisches Erkennen*, D. Rowe, ed. Basel, Switzerland, Birkhäuser.

Arezoo Islami (2016). A Match Not Made in Heaven: On the Applicability of Mathematics in Physics. *Synthese*. vol. 193, no. 7, 1–23. doi: 10.1007/s11229-016-1171-4.

Israel Kleiner (2007). *A History of Abstract Algebra*. Basel, Switzerland: Birkhäuser.

Helge Kragh (1990). *Dirac: A Scientific Biography*. Cambridge, U.K.: Cambridge Univ. Press.

T. S. Kuhn (1963). *Arch. for Hist. of Quantum Physics*, Interview with Eugene P. Wigner, Inventory of Microfilms of Manuscript Materials. Joan N. Warnow Center for History of Physics, American Institute of Physics. 1986. See https://www.aip.org/history-programs/niels-bohr-library/oral-histories.

Jesper Lützen (2011). The Physical Origin of Physically Useful Mathematics, *Interdiscip. Science Rev.* 36:3, 229–243.

Paolo Mancosu (2010). *The Adventure of Reason*. Oxford, U.K.: Oxford Univ. Press.

Herbert Mehrtens (1990). *Moderne—Sprache—Mathematik*. Suhrkamp, Frankfurt/Main, Germany.

Gregory H. Moore (1982). *Zermelo's Axiom of Choice*. Berlin: Springer.

Paul J. Nahin (1998). *An Imaginary Tale: The Story of $\sqrt{-1}$*, Princeton, NJ: Princeton Univ. Press.

John von Neumann (1947). The Mathematician, in *Works of the Mind*, vol. I, no. 1 (Chicago: Univ. of Chicago Press, 1947), 180–196. Also in von Neumann's *Collected Works*.

Wolfgang Pauli (1926). On the hydrogen spectrum from the standpoint of the new quantum mechanics. *Z. Physik* 36, 336–363. Reprinted in B. L. Van der Waerden, *Sources of Quantum Mechanics* (Dover, New York, 1968).

Henri Poincaré (1905). *The Value of Science*. New York: Science Press, 1907 (English ed.).

Bernhard Riemann (1854). Über die Darstellbarkeit einer Function durch eine trigonometrische Reihe (Habilitationsschrift), *Abhandlungen der Königlichen Gesellschaft der Wissenschaften zu Göttingen* **13** (1868). In Riemann, *Gesammelte mathematische Werke und wissenschaftlicher Nachlass*, ed. H. Weber (Leipzig, Germany: Teubner, 1892), 227–265.

David Rowe (2000). The Calm Before the Storm: Hilbert's Early Views on Foundations. In: *Proof Theory: History and Philosophical Significance*, V. F. Hendricks, S. A. Pedersen, K. F. Jørgensen, eds. Dordrecht, Netherlands: Springer-Science, pp. 55–93 (Synthese Library, vol. 292).

Steve Russ (2011). The Unreasonable Effectiveness of Mathematics in the Natural Sciences. *Interdiscip. Sci. Rev.* 36:3, 209–213.

Erhard Scholz (2006). Introducing groups into quantum theory (1926–1930). *Historia Mathematica* 33, 440–490.

Issai Schur (1905). Neue Begründung der Theorie der Gruppencharaktere. (New foundation for the theory of group characters) *Sitzungsberichte der Akademie der Wissenschaften zu Berlin*, 406–432. Reprinted in Schur, *Gesammelte Abhandlungen* vol. 1, Springer, 143–169.

Andrew Szanton (1992). *The Recollections of Eugene P. Wigner*, New York: Springer US (Plenum Press, 1992).

Hermann Weyl (1928). *The Theory of Groups and Quantum Mechanics*. New York: Dover, 1931.

Eugen Wigner (1931). *Gruppentheorie und ihre Anwendungen auf die Quantenmechanik der Atomspektren*, Braunschweig, Germany: Vieweg Verlag. (English ed.) *Group Theory and Its Application to the Quantum Mechanics of Atomic Spectra*. New York: Academic Press, 1959.

—— (1960). The unreasonable effectiveness of mathematics in the natural sciences. Also in Wigner (1967) and in *Coll. Works*, vol. 6 (1995), 534–559; we give page references to the first ed. and the 1995 *Works*.

—— (1967). *Symmetries and Reflections: Scientific Essays*. Bloomington, IN: Indiana Univ. Press, 1970, Cambridge, MA: MIT Press.

—— (1995). *The Collected Works of E. Wigner*, Part B (Historical, Philosophical, and Socio-Political Papers) Jagdish Mehra, ed. vol. 6: *Philosophical Reflections and Syntheses*. Berlin, New York: Springer.

Learning and Teaching Interdisciplinary Modeling

CHRIS ARNEY

Interdisciplinary modeling combines concepts, methods, techniques, and elements of various disciplines (in the sciences, humanities, and arts) to

- obtain solutions to problems;
- develop understanding of issues;
- provide recommendations to decision makers; and
- implement and build tools, algorithms, and systems.

To be effective for society, interdisciplinary modeling must provide the capability for analysts to solve realistic and challenging problems. Good education programs teach students both disciplinary and interdisciplinary modeling and problem-solving methods, and they provide opportunities for students to practice and hone their modeling skills. The Interdisciplinary Contest in Modeling (ICM) experience is one way to build experience and refine skills.

Here we look at the nature, processes, education, and resources related to interdisciplinary modeling and problem solving, with the hope that students can use this information to prepare for the ICM and improve their interdisciplinary modeling skills.[1]

Interdisciplinary Problems

Real issues and modern problems can have many challenging characteristics. Some of these are the following:

- Intransparency (lack of clarity of the situation or changing environments and criteria);

- Multiple goals (many stakeholders with competing criteria);
- Complexity (large numbers of items, interrelations, decision elements, dimensions, geometries, and time scales);
- Dynamics (time considerations, constraints, and sensitivities);
- Spatial and geometric considerations (integral or fractional dimensions); and
- Political and social elements (human or cyber considerations).

One element to avoid or minimize in modeling is *confirmation bias*, which is favoring a preconceived notion. Confirmation bias can dramatically harm or constrain modeling and problem solving. Modelers must be aware of and adapt their models to avoid or resist irrelevant, biased, or erroneous information. Since data are never perfectly accurate nor completely clean, considerable effort to reduce errors or eliminate bad data is needed. ICM problems often require data to be considered—and sometimes obtained or generated—by the teams. This collection, choosing, and weighing of data are important steps in the modeling process that should not be treated lightly by the teams.

Mathematical Modeling

Mathematical modeling is a structured process with many loops and choices that can make it as much art as science. In performing this process, the modeler needs to describe the phenomena in mathematical terms. The four basic steps in the process (as described in Arney [2014, 169–170]) are the following:

- **STEP 1: IDENTIFY THE PROBLEM.** The problem is stated in as precise a form as possible. Sometimes, this is an easy step; other times this may be the most difficult step of the entire process.
- **STEP 2: DEVELOP A MODEL.** This is both a translation from the natural language statement made in Step 1 to mathematical language but also the development of relationships between the factors involved in the problem. Because real-world situations are often too complex to allow the modeler to account for every facet of the situation, simplifying assumptions must be made. Data collection is often part of model construction. Variables are defined, notation is established, and some form of mathematical relationship and/or structure is established.

- **STEP 3: SOLVE THE MODEL.** The model is solved so that the answer is understood in the context of the original problem. If the model cannot be solved, it may need to be simplified by adding more assumptions in Step 2.
- **STEP 4: VERIFY, INTERPRET, AND USE THE MODEL.** Before using the model, it should be tested or verified that it makes sense and works properly. Its output should be interpreted in the context of the problem. It is possible that the model works, but it's too cumbersome or too expensive to implement. The modeler returns to earlier steps to adjust as needed.

The modeling process is iterative in the sense that the modeler may need to go back to earlier steps and repeat the process or continue to cycle through the entire process (or part of it) several times. If the model cannot be solved or is too cumbersome to use, the model is simplified. If the model needs to be more powerful, or more complication or rigor needs to be added, the process of relaxing assumptions is called *refining* the model. By simplifying and refining, the modeler can adjust the realism, accuracy, precision, and robustness of the model. By using this mathematical modeling process, modeling students can gain confidence to approach complex and difficult problems and even develop their own innovative approaches to solving problems.

Interdisciplinary Modeling

Interdisciplinary modeling is a creative process that, while sometimes based on structured processes such as mathematical modeling, usually involves an innovative and complex combination of modeling and problem-solving methods from various disciplines and schools of thought.

The traditional modeling process was based on making viable and appropriate assumptions and connections to produce a framework. This structured, Newtonian style of modeling and problem solving was often based in mathematics, mechanics, engineering, and physical science (Teller [1980]).

With the advent of the computer and the availability of tremendous amounts of data, modern interdisciplinary modeling often reduces assumptions to a minimum and attempts to embrace the complexity of the real situation. Interdisciplinary modeling combines established methodologies with novel procedures in its processes and structures, thus allowing for complexity and specificity in its framework. The model is then solved, used, implemented, tested, and/or validated, to

- produce a measure,
- design an algorithm,
- solve a problem,
- accomplish a task,
- understand a phenomenon,
- build a system, and/or
- make a decision.

Modeling can and usually does rely on research that incorporates accurate scientific information and data, relevant knowledge, and innovative perspectives in the model. Interdisciplinary modeling can be quantitative or qualitative, but most viable modern models are hybrid and incorporate many different kinds of steps and processes.

The modern form of interdisciplinary modeling is called by various names, such as *network modeling, data science, operations research, analytics, informatics*, and *information science.*

The most emblematic inventor of these kinds of processes was mathematician Norbert Wiener, whose theory of cybernetics included models with iterative control and feedback loops. After the initial stages of cybernetics in the 1950s, interdisciplinary modeling was used in design and analysis of communication systems, electronics, biological systems, and economics (Wiener [1961]).

This method of interdisciplinary modeling was the key to unlocking issues in computing, artificial intelligence, neurobiology, psychology, and sociological systems. This new modeling paradigm not only allowed for the entry of the fields of life, behavioral, and social sciences into the modeling world, but also began to challenge the Newtonian simplicity assumption at a conceptual level. Other interdisciplinary modelers soon followed to make interdisciplinary modeling a highly valuable methodology and tool. Lorenz provided rigorous backing through chaos theory and strange attractors to show that even a perfectly deterministic system can behave in erratic ways. Benôit Mandelbrot [1977] demonstrated that a high level of complexity exists in rudimentary geometric objects that make up the world as we know it. These ideas are now conceptual elements of interdisciplinary modeling.

Measures and Metrics

Building good measures for system properties, data, and the sensitivities of their effects on the achievement of goals for the problem are important, especially if the data set is wider than it is deep, thereby affecting most of the elements of the model. Good measures are needed in many models, especially when the problem is quantitative. In qualitative modeling, the measuring is often performed by comparison. Marcus Weeks [2010] discusses determination of size by a comparison methodology and makes the size comparison relevant to humans.

A Course in Modeling

Interdisciplinary modeling courses seek to address the complex process of translating real-world events into mathematical and scientific language, solving or running the resulting model (iterating as necessary), and interpreting the results in terms of real-world issues. Topics often include model development from data, regression, general curve fitting, and deterministic and stochastic model development. Easley and Kleinberg [2010] and Newman [2010] are textbooks that are helpful in such a course. *Guidelines for Assessment and Instruction in Mathematical Modeling Education* (GAIMME) edited by Garfunkel and Montgomery [2016] is an excellent reference for teaching foundational courses. There are many good modeling textbooks, and some are listed below.

Interdisciplinary projects based on actual problems and issues are used to integrate the various topics of modeling for the student. Through such a course, students should be able to

- Frame a question using mathematics and science to begin developing a model,
- Understand different modeling approaches and the trade-offs with selecting an approach, and
- Interpret the results of a model.

During a section on optimization, students learn to

- Take a scenario and transform it into a model focused on optimization;
- Understand how to set up an optimization formulation with decisions, an objective, and constraints;

- Understand the unique aspects of certain optimization cases, such as linear, integer, and dynamic programming;
- Take the solution to an optimization formulation and interpret the answer; and
- Understand the robustness and sensitivity of a model.

Through the study of dynamical systems modeling, students learn to

- Take a scenario and transform it into a model focused on a dynamical system;
- Understand how to set up a dynamical system with an independent variable, dependent variables, and a governing relationship;
- Interpret the model without solving the differential equations;
- Understand the robustness of the model (sensitivity); and
- Understand the possibility of chaotic solutions.

From a stochastic point of view, students learn to

- Take a scenario and transform it into a model while incorporating uncertainty;
- Understand how to set up a probabilistic model with a random variable, sample space, and distribution; and
- Understand how to set up a Markov model with a state space, a random variable with distribution, and state transitions.

Future Trends in Modeling

Society and organizations need experts in interdisciplinary modeling in order to make better and faster decisions. Future modelers will need to build viable models to confront complex multidisciplinary and interdisciplinary issues in our information-centric world. High-impact areas include analyzing many issues for a larger and smarter Internet, automation of knowledge through artificial intelligence and machine learning, high-powered cloud computing, autonomous vehicles, and smart robots. Modeling in these areas is by nature interdisciplinary.

A striking example of a project that reveals the future of interdisciplinary modeling is IBM's development of the Watson system to compete in the information-centric game Jeopardy. Watson's notable success in that first endeavor, and in many applications since, illustrates the potential of interdisciplinary modeling. Modern science is embracing the future contributions of interdisciplinary modeling as science shifts from little science (single investigators) and big science (large labs working on specific projects) to a future of team science (multidisciplinary teams of scientists, much like the IBM Watson team and the ICM team framework) (West [2016]). Modern science, through its use of interdisciplinary modeling, is building a collective power that is more creative, more original, and more effective than any single disciplinary perspective and the simplicity-focused Newtonian modeling of the past.

Another example of interdisciplinary modeling is found in cyberspace, where computing and networking are important elements in the model, but so are ethical, political, and social elements. Data science, human psychology, and many other disciplines are all parts of the virtual and digital cyberworld. The complex issues in cybermodeling are twofold:

- What is the balance between security and performance versus privacy and information availability?
- How do models treat the underlying competitive nature of the attacker versus defender dynamic?

Hackers and malicious systems are pitted against defenders of freedom, information, and the system's performance. These game-theoretic settings take interdisciplinary modeling to new heights of what-if, cause–effect, and who-did-it questions. Cyberproblem solving can require high-dimensional, nonlinear models, with dynamic structures and processes to adapt to the changing situations. A major challenge is that the same elements of the network that create its positive attributes (effectiveness and freedom) also produce its negative elements (vulnerability and lack of privacy and security). What makes a network robust, survivable, and hard to kill, paradoxically can also make it inefficient, difficult to manage, and vulnerable to penetration.

Diversity is often the model's attribute that best provides the potential for resilience to vulnerabilities and yet limits agility. One natural way to create diversity in cybersystems is through randomness (explicitly designed random processes). As a result, future interdisciplinary modelers will need to build diversity and randomness into many of their models and systems.

The Interdisciplinary Contest in Modeling (ICM)

The ICM tries to mimic the elements of real-life problem solving previously outlined. Real-life problem solving is inherently interdisciplinary, and therefore education programs should include interdisciplinary modeling. *Policy modeling* has become a popular way to inform decision makers of potential priorities and determine the what-if effects of different scenarios or decisions; so we intend to continue that type of modeling problem in the ICM. A more traditional field where interdisciplinary modeling plays a major role is *environmental science*. And, since interdisciplinary modeling can include many types of knowledge and perspectives, we intend to continue problems in *operations research and network science* that bring modern issues to the ICM problem set. So we will continue problems in the three current areas of the ICM: operations research and network science, environment, and policy.

Quite often, the assembly and pathways from which interdisciplinary models are built or implemented are artistic and lead to qualitative models. Good problem solving thus involves making necessary assumptions and adding appropriate complexity that leads to appropriately quantitative or qualitative models using scientific and artistic methods. Many viable models are hybrid—containing quantitative, qualitative, scientific, and artistic elements—and incorporate many different disciplinary and interdisciplinary structures and processes.

Resources

The following three lists of problem materials can help interdisciplinary modelers learn methods, see examples, and gain understanding of the modeling process. The lists are divided into

- modeling and problem-solving books and articles,
- interdisciplinary books and articles, and
- journals relevant to interdisciplinary modeling.

Modeling and Problem-Solving Books and Articles

Albright, Brian. 2010. *Mathematical Modeling with Excel*. Burlington, MA: Jones and Bartlett.

Beckmann, J. F., and J. Guthke. 1995. Complex problem solving, intelligence, and learning ability. In Frensch and Funke [1995], 177–200.

Bender, Edward A. 1978. *An Introduction to Mathematical Modeling.* New York: Wiley. 2012. Reprint. Mineola, NY: Dover.

COMAP. 2012. *The Mathematical Modeling Handbook.* Print and CD-ROM. Bedford, MA: COMAP.

———. 2013. *The Mathematical Modeling Handbook II: The Assessments.* Print and CD-ROM. Bedford, MA: COMAP.

———. 2015. *The Mathematical Modeling Handbook III: Lesson Paradigms.* CD-ROM. Bedford, MA: COMAP.

Dym, Clive L. 2004. *Principles of Mathematical Modeling.* 2nd ed. New York: Academic Press.

Farid, Mohammed M. 2010. *Mathematical Modeling of Food Processing.* Boca Raton, FL: CRC Press.

Ford, Andrew. 2009. *Modeling the Environment.* 2nd ed. Washington, DC: Island Press.

Frensch, P. A., and J. Funke (eds.). 1995. *Complex Problem Solving: The European Perspective.* Hillsdale, NJ: Lawrence Erlbaum Associates.

Giordano, Frank R., William P. Fox, and Steven B. Horton. 2013. *A First Course in Mathematical Modeling.* 5th ed. Boston: Brooks-Cole.

Jones, Beau Fly, Claudette M. Rasmussen, and Mary C. Moffitt. 1997. *Real-Life Problem Solving: A Collaborative Approach to Interdisciplinary Learning.* Washington, DC: American Psychological Association.

Meerschaert, Mark M. 2013. *Mathematical Modeling.* 4th ed. New York: Academic Press.

Meyer, Walter J. 1984. *Concepts of Mathematical Modeling.* New York: McGraw Hill. 2004. Reprint. Mineola, NY: Dover.

Otto, Sarah, and Troy Day. 2007. *A Biologist's Guide to Mathematical Modeling in Ecology and Evolution.* Princeton, NJ: Princeton University Press.

Polya, George. 1945. *How to Solve It: A New Aspect of Mathematical Method.* Princeton, NJ: Princeton University Press. 1957. 2nd ed. New York: Doubleday. 2004. Reprinted with foreword by John Conway. Princeton, NJ: Princeton University Press. 2009. Reprinted with foreword by Sam Sloan. San Rafael, CA: Ishi Press.

Schoenfeld, A. H. 1985. *Mathematical Problem Solving.* Orlando, FL: Academic Press.

Shier, Douglas R., and K. T. Wallenius. 1999. *Applied Mathematical Modeling: A Multidisciplinary Approach.* Boca Raton, FL: CRC Press.

Sokolowski, John A., and Catherine M. Banks. 2009. *Principles of Modeling and Simulation: A Multidisciplinary Approach.* Hoboken, NJ: Wiley.

———. 2009. *Modeling and Simulation for Analyzing Global Events.* Hoboken, NJ: Wiley.

Walloth, Christian, Jens Martin Gurr, and J. Alexander Schmidt. 2014. *Understanding Complex Urban Systems: Multidisciplinary Approaches to Modeling.* New York: Springer.

Xin-She Yang. 2013. *Mathematical Modeling with Multidisciplinary Applications.* New York: Wiley.

INTERDISCIPLINARY BOOKS AND ARTICLES

Augsburg, Tanya. 2005. *Becoming Interdisciplinary: An Introduction to Interdisciplinary Studies.* Dubuque, IA: Kendall/Hunt.

Davies, M., and M. Devlin. 2007. *Interdisciplinary Higher Education: Implications for Teaching and Learning.* Melbourne, Australia: Centre for the Study of Higher Education, University of Melbourne.

Dluhy, Milan J., and Kan Chen (eds.). 1986. *Interdisciplinary Planning: A Perspective for the Future.* New Brunswick, NJ: Center for Urban Policy Research. 2012. Reprint. New Brunswick, NJ: Transaction Publishers.

Frodeman, Robert (ed.). 2010. *The Oxford Handbook of Interdisciplinarity.* New York: Oxford University Press.

Frodeman, R., and C. Mitcham. 2007. New directions in interdisciplinarity: Broad, deep, and critical. *Bulletin of Science, Technology, and Society* 27 (6): 506–514.

Hendler, James, Nigel Shadbolt, Wendy Hall, Tim Berners-Lee, and Daniel Weitzner. 2008. Web science: An interdisciplinary approach to understanding the Web. *Communications of the Association for Computing Machinery* 51 (7): 60–69.

Klein, Julie Thompson (ed.). 2002. *Interdisciplinary Education in K-12 and College: A Foundation for K-16 Dialogue.* New York: The College Board.

———. 2005. *Humanities, Culture, and Interdisciplinarity: The Changing American Academy.* Albany NY: State University of New York Press.

———. 2006. Resources for interdisciplinary studies. *Change: The Magazine of Higher Learning* 38 (2) (March/April): 50–56.

———. 2010. *Creating Interdisciplinary Campus Cultures: A Model for Strength and Sustainability.* San Francisco, CA: Jossey-Bass.

Kleinberg, Ethan. 2008. Interdisciplinary studies at the crossroads. *Liberal Education* 94 (1): 6–11.

JOURNALS RELEVANT TO INTERDISCIPLINARY MODELING

Applied Mathematical Modeling
Chaos: Interdisciplinary Journal of Nonlinear Science
International Journal of Modeling, Simulation, and Scientific Computing
Journal of Informatics
Journal of Interdisciplinary Modeling and Simulation
Journal of Policy Modeling
Math and Computer Education Journal
Mathematical Modeling and Numerical Analysis
Mathematical Models and Methods in Applied Science
PRIMUS: Problems, Resources, and Issues in Mathematics Undergraduate Studies
SIAM: Multiscale Modeling and Simulation
The UMAP Journal of Undergraduate Mathematics and Its Applications

Conclusion

Interdisciplinary modeling seeks to connect the various problem-solving methodologies and perspectives that exist across many disciplines. To be effective for society, modeling must provide the capability for analysts to solve modern realistic and challenging problems. The ICM experience is one way to build experience and refine interdisciplinary modeling skills.

Many good educational programs offer students both disciplinary and interdisciplinary problem-solving opportunities, but often more modeling and problem solving are needed outside the classroom in order to develop analysts. This is where the ICM plays an important role. The problem categories for next year's ICM will be the same as the last two years: operations research and network science, the environment, and policy.

Note

1. The opinions in this article are the author's alone and do not necessarily reflect the opinion of his colleagues, the U.S. Military Academy (West Point), the Department of the Army, or any other U.S. government agency.

References

Arney, Chris. 2014. Developing and understanding interdisciplinary problem solving. In *ICM Interdisciplinary Contest in Modeling: Culturing Interdisciplinary Problem Solving*, edited by Chris Arney and Paul J. Campbell, 165–176. Bedford, MA: COMAP.

Easley, D., and J. Kleinberg. 2010. *Networks, Crowds, and Markets: Reasoning about a Highly Connected World*. New York: Cambridge University Press.

Garfunkel, Sol, and Michelle Montgomery. 2016. *Guidelines for Assessment and Instruction in Mathematical Modeling Education*. Bedford, MA: COMAP.

Lorenz, Edward. 2005. Designing chaotic models. *Journal of the Atmospheric Sciences* 62 (5) 1574–1587.

Mandelbrot, Benôit B. 1977. *The Fractal Geometry of Nature*. San Francisco: W. H. Freeman and Company.

Newman, Mark. 2010. *Networks: An Introduction*. New York: Oxford University Press.

Teller, Edward. 1980. *The Pursuit of Simplicity*. Malibu, CA: Pepperdine University Press.

West, Bruce. 2016. *Fractional Calculus View of Complexity: Tomorrow's Science*. Boca Raton, FL: CRC Press.

Weeks, Marcus. 2010. *How Many Elephants in a Blue Whale? Discover a Whole New Way to Measure the World*. Lewes, UK: Ivy Press.

Wiener, Norbert. 1961. *Cybernetics, or Control and Communication in the Animal and the Machine*. Cambridge, MA: MIT Press.

Six Essential Questions
for Problem Solving

NANCY EMERSON KRESS

Martina joined my precalculus class in November of her senior year of high school. (The name and some details of this student's experience have been changed to protect her identity.) She told me that she did not consider herself to be good at math, but she felt that a good grade in precalculus would be an important part of her college applications. She also told me that college was critical to her ability to live a better life than she had had as a child and that she was certain she would be able to do well in precalculus as long as I was very clear about exactly what I wanted her to do.

One of the primary expectations I have for my students is for them to develop greater independence when solving complex and unique mathematical problems. Martina joined the majority of the class in telling me that she was comfortable with explicit, step-by-step instructions, but that working independently on math problems that were different from those she had previously been shown how to solve was very difficult. Martina was especially emphatic about the nature and degree of this challenge. The expectations that many of my students had of me with regard to prescriptive, step-by-step instructions were at odds with my expectations for them—and yet we avoided conflict. The story of how I supported my students as they gained confidence and independence with complex and unique problem-solving tasks, while honoring their expectations with regard to clear, explicit instruction, is rooted in a set of guiding questions I call *essential questions for problem solving*.

Why We Use Essential Questions for Problem Solving

One challenge that teachers face is developing instructional methods that support the continued growth of successful problem solvers and simultaneously nurturing the development of confidence and enabling success among students who struggle. It is incumbent on the mathematics teaching community to implement teaching strategies that genuinely reflect the belief that all students are able to learn and do mathematics (Boaler 2016; Dweck 2007) and that provide every student with robust support and rich opportunities to expand their problem-solving skill set.

A strategy commonly used to improve students' success with problem solving is to increase their exposure to challenging and interesting problems. Providing support and teaching students that the key to success is perseverance (Boaler 2016; Dweck 2007) may increase their ability to solve a variety of problems in the future. However, experience and perseverance alone do not consistently lead all students to become successful with complex or unique problem-solving tasks.

Students are not all equally prepared to participate in open curricular and reform approaches to learning mathematics (Boaler 2002). There is concern that some students may be less aware of the particular mathematical practices that are being used and developed in their classes (Ball et al. 2005; Boaler 2002; Lubienski 2000; Selling 2016), and for many this is because they expect teaching to be more direct (Delpit 1988). This concern is especially strong in relation to students from lower socioeconomic status or working-class backgrounds, students who speak English as a second language, and students who belong to minority racial or ethnic groups (Ball et al. 2005; Boaler 2002; Delpit 1988; Lubienski 2000; Parks 2010).

The term *explicit*, as applied to mathematics teaching, is often associated with step-by-step procedural instruction. This is the form of instruction my students were most comfortable with. But Boaler (2002) cautions against responding to equity concerns with a return to more direct teaching methods because direct instruction can reduce students' opportunities to engage in sense making about complex problems (Boaler 2002; Greeno and Boaler 2000; Schoenfeld 1992; Selling 2016).

Improving equity, particularly for students who respond positively to direct teaching methods, without resorting to prescriptive methods of teaching requires a more nuanced understanding of what it means to be explicit. Selling (2016) suggests an alternative to being explicit that centers on bringing direct attention to mathematical practices being used in the classroom. She claims that "participants in this interaction may be more or less aware that they (or others) are engaging in particular mathematical practices" (p. 510). She suggests that a form of being explicit that is direct about highlighting mathematical practices, as opposed to step-by-step instructions, enables all students to be equally aware of the strategies and practices being used. This interpretation has important implications for increasing equitable access to mathematics for students from widely varied backgrounds.

Essential Questions as a Framework

The essential questions described in this article are designed to provide a framework to support students to pose purposeful questions (NCTM 2014, p. 35) about complex mathematical problems, and they are consistent with design principles for active learning (Webb 2016). Although these questions differ significantly from the questions and suggestions that Pólya proposed (1945), they take a similar approach in the sense that they are applicable to a wide range of problem-solving tasks and do not prescribe specific mathematical steps for solving a particular type of problem. I used these questions in second-year algebra and precalculus classes to support increased learning opportunities for all students.

The first step in developing essential questions for problem solving was to identify a specific skill set that would support reliable and consistent success at problem solving for all students. It involved observing precisely which actions students were taking that lead to success, regardless of whether students themselves consciously identified the critical practices they were using. The actions and skills identified as both fundamental and comprehensive for supporting problem-solving success across a wide range of problem types are as follows:

- Noticing, or making observations;
- Asking questions;

- Knowing how and why to carry out particular mathematical actions; and
- Verifying accuracy.

Desirable strategies that are less frequently applied, even among highly successful students include these:

- Making connections and
- Extending the problem.

A study of the alignment of the questions to the Common Core's (CCSSI 2010) Standards for Mathematical Practice (SMP) was carried out (Kress 2014), and the questions were refined over the course of two years of application in precalculus and second-year algebra.

How to Introduce Six Essential Questions for Problem Solving

When I first introduce these questions to a class, I use a prompt—just a single quadratic function—that is not complex or unique at all. The task's lack of complexity allows me to introduce the questions as a structure to support students' exploration of mathematics. They can make observations, ask for additional information, and try out ideas. After one fifty-minute class period, students see how previously isolated topics fit together to form a big picture. Students seem to gain satisfaction from the experience of making multiple connections between concepts that they have previously experienced as isolated, and use of the questions increases both the depth and breadth of students' understanding of quadratic functions and their graphs.

The six questions are listed below, followed by a description of the purpose and role of each question, as well as examples of how my students took up and responded to the questions.

1. What do you notice?
2. What additional information or clarification would be helpful?
3. What can you do or figure out?
4. How do you know that your work and/or answer are accurate?
5. Is there another way you could approach this problem?
6. What else can you say about the problem, and what else would you like to know?

What Do You Notice?

Many students immediately attempt to begin mathematical work on problems without pausing to consider the overall picture or subtle details. This first question supports students in taking stock of what they know before they get embroiled in the complexities of the task. This question also provides the teacher with formative assessment information.

When Martina and her precalculus classmates were asked to consider a single quadratic function as a prompt and were asked, "What do you notice?" the responses included observations such as these:

- "There's a little two above the x."
- "It's a quadratic."
- "You could factor."
- "There are three terms."
- "You could graph it."
- "I think the graph might be a parabola."

Second-year algebra students who had been introduced to quadratic functions in their first algebra course responded similarly. I made certain that every student response was accepted and publicly recorded.

What Additional Information or Clarification Would Be Helpful?

Some students hesitate to ask questions. Others ask for help without putting effort into refining their questions or identifying what aspect of the problem requires clarification. This prompt legitimizes students asking for additional information while supporting student ownership and responsibility for the thought process.

When I asked students to consider a quadratic function and, "What additional information or clarification would be helpful?" they responded with questions such as the following:

- "What does the two above the x mean?"
- "How would you factor that?"
- "What does the graph look like?"
- "How do you know it's a quadratic?"

Questions can be answered immediately, either by student volunteers or by the teacher. If they are not answered immediately, then establishing a strategy to ensure that all questions are answered in the course of that class period is imperative.

WHAT CAN YOU DO OR FIGURE OUT?

This is the point at which students do the work of attempting to determine an answer if the prompt calls for one. In the context of a quadratic function prompt, my students typically explore the mathematics and draw connections and conclusions.

In second-year algebra and precalculus, my students did some or all of the following:

- Factored,
- Made a table of values,
- Drew a graph,
- Solved for x-intercepts using the quadratic formula,
- Stated x-intercepts,
- Stated the y-intercept, and
- Stated the coordinates of the vertex.

Letting students determine the direction of the discussion is important. Students usually think of new ideas that build off one another's responses, but the order in which this work happens varies from class to class. I facilitate the discussion by calling on students and taking detailed notes on the board. Another teacher might opt to ask a student to take the notes on the board. It is important that the notes are visible to everyone as the students work together. By the end of this stage, my classes have created a board covered with mathematical calculations and information related to the quadratic function on which they are working.

HOW DO YOU KNOW THAT YOUR WORK AND/OR ANSWER ARE ACCURATE?

Students have many different methods of verifying accuracy. One is to solve the problem in a different way, confirming that the same result is obtained. Students may also consider whether their observations contain contradictions, substitute an answer back into an equation to verify its validity, or check that the solution obtained from an equation agrees with what is shown on a graph. Methods vary greatly by student as well as by type of problem.

Martina and her classmates, when applying this question to the quadratic function $f(x) = x^2 + 3x - 4$, responded in the following ways:

- "The y-intercept on the graph is at -4, and in the equation, when you make $x = 0$, then $y = -4$."
- "The x-intercepts are at 1 and -4, and the factors are $(x + 4)$ and $(x - 1)$."
- "The graph opens up, and the coefficient of x^2 is positive."
- "The graph is symmetrical, so it looks right."

If some of their work contains an error, they may observe the following:

- "We drew the graph with a y-intercept at -3, but the constant in the equation is -4. That doesn't make sense."
- "Our graph crosses the x-axis at 4, but when I substitute 4 into the function, I don't get 0 for an answer."

If students are slow to respond, follow-up questions are necessary. I used such probing questions as these:

- "How are the y-intercept and constant in the function related to each other?"
- "What do you notice if you substitute 4 into the function for x?"

Is There Another Way You Could Approach This Problem?

Students often address this question in conjunction with the previous one because solving the problem in another way is an efficient method of confirming the accuracy of their work. They are presented separately because both topics are of significant importance. Because they can be answered separately, considering them as unique concerns is valuable.

This consideration does not always result in a second practical method of solving the problem, but even when it does not, students typically gain additional insights through investigating the possibility of other angles.

When considering a quadratic function, this is the point at which my students are likely to notice if they have omitted the use of a familiar strategy, such as the quadratic formula or factoring. They add that work to what they have done previously, further strengthening their ability to confirm that their previous observations make sense.

What Else Can You Say About the Problem, and What Else Would You Like To Know?

These questions serve the purpose of prompting students to stop and think before moving on. Having arrived at a solution and confirmed that the process and answer are accurate, students are frequently prepared to be finished with a task they feel has been completed. This question encourages students to reflect on the ways in which the work fits into their larger experience or general knowledge base. It also provides additional assessment information to the teacher.

Students in my second-year algebra and precalculus classes asked questions and made new observations such as these:

- "What would make the graph open down?"
- "Can a parabola open sideways?"
- "What would the graph look like if there were a different coefficient for x^2?"
- "The $+/-$ in the quadratic formula gives you values to the right and left of the vertex, or the axis of symmetry. That's how you get the x-intercepts. You could even write it as this:

$$\frac{-b \pm \sqrt{b^2 - 4ac}}{2a}.\text{"}$$

Teaching students to use these essential questions purposefully as prompts to work through stages of solving problems increased engagement in my classes. The questions shaped and directed students' thinking and supported all students in becoming aware of the use of these practices for solving mathematical problems. Students' confidence increased, and they began doing work that was more thorough and complete.

Next Steps: Moving toward Independence

When Martina first joined my class, she struggled to participate within the existing norms of the classroom. When the class engaged in discourse about mathematics, collective sense making around open-ended problems, or communal exploration of multiple methods of solution, she did not participate. If I did not provide immediate explanations of step-by-step procedures for solving problems, she got frustrated and stopped participating for the rest of the class period.

The structure of the essential questions for problem solving supported Martina to reduce her dependence on procedural instruction. Her willingness to engage in open-ended and complex tasks gradually increased, and she became more likely to participate in class and group discussions. She developed the ability to generate observations, ideas, and solution strategies independently. She and her classmates built on their experience working with a simple quadratic function prompt, and they became more comfortable working on complex problems such as the following:

> To celebrate the Fourth of July, a city has hired Star Burst, Inc. to launch fireworks into the air from the top of a tower 20 feet tall. The fireworks can be fired with an initial upward velocity of 128 feet per second. Write a mathematical model for this scenario, and use your model to find how many seconds after launch the fireworks attains its maximum height (assuming it has not yet exploded). What is its height above ground at this time? Explain how you know that this is the maximum height. If you wanted the fireworks to reach its maximum height exactly 5 seconds after launching, how might you accomplish this?

Early in the school year, my students would have responded to the problem described above by asking for demonstration of step-by-step strategies using a nearly identical problem. Later in the year, Martina and her classmates were able to read such a problem and use the essential questions to work their way through the scenario. They developed greater ability to engage in productive struggle (NCTM 2014, pp. 48–52), and they demonstrated the ability to persist through a process to solve problems unlike those they had seen before.

I found that the essential questions proved useful with a variety of concepts and problem types, including—but not limited to—graphing rational, higher-order polynomial, trigonometric, exponential, and logarithmic functions as well as a variety of modeling scenarios.

In Conclusion

Teaching problem-solving skills equitably is challenging. Detailed procedural instruction supports students in the short term but runs the risk of undermining students' independence. Being explicit about mathematical practices that lead to effective problem solving has the potential to increase equitable access to high-level learning about complex mathematical problem-solving tasks. Essential questions for problem solving differentiate and personalize instruction by providing structure for students who need it, while helping successful students to recognize and take ownership of the actions underlying their success.

Acknowledgments

The author wishes to acknowledge the contributions of Laura Taylor Kinnel, who read and provided valuable feedback on an early draft of this article, and one anonymous reviewer, whose persistence and advice were instrumental in the article becoming what is published here. The author also acknowledges the support of National Science Foundation (NSF) award no. 1624610. Any opinions, findings, and conclusions or recommendations expressed in this material are those of the author and do not necessarily reflect the views of NSF.

Bibliography

Common Core State Standards Initiative (CCSSI). 2010. Common Core State Standards for Mathematics. Washington, DC: National Governors Association Center for Best Practices and the Council of Chief State School Officers, http://www.corestandards.org/wp-content/uploads/Math_Standards.pdf

Ball, Deborah Loewenberg, Imani Masters Goffney, and Hyman Bass. 2005. "The Role of Mathematics Instruction in Building a Socially Just and Diverse Democracy." *The Mathematics Educator* 15 (1): 2–6.

Boaler, Jo. 2002. "Learning from Teaching: Exploring the Relationship between Reform Curriculum and Equity." *Journal for Research in Mathematics Education* 33, no. 4 (July): 239–58. https://doi.org/10.2307/749740.

————. 2016. *Mathematical Mindsets*. San Francisco: Jossey-Bass.

Common Core State Standards Initiative (CCSSI). 2010. "Common Core State Standards for Mathematics." Washington, DC: National Governors Association Center for Best Practices and the Council of Chief State School Officers, http://www.corestandards.org/wp-content/uploads/Math_Standards.pdf.

Delpit, Lisa D. 1988. "The Silenced Dialogue: Power and Pedagogy in Educating Other People's Children." *Harvard Educational Review* 58 (3): 280–98.

Dweck, Carol S. 2007. *Mindset: The New Psychology of Success*. New York: Ballantine Books.

Greeno, James G., and Jo Boaler. 2000. "Identity, Agency, and Knowing in Mathematics Worlds." In *Multiple Perspectives on Mathematics Teaching and Learning*, edited by Jo Boaler, pp. 171–200, Book 1, International Perspectives on Mathematics Education Series. Westport, CT: Ablex Publishing.

Kress, Nancy Emerson. 2014. "Essential Questions for Mathematics: Developing Confident, Knowledgeable, Creative Students." Paper presented at the New Hampshire Teachers of Mathematics Spring Conference, NHTI, Concord's Community College, Concord, NH, March 17.

Lubienski, Sarah Theule. 2000. "Problem Solving as a Means toward Mathematics for All: An Exploratory Look through a Class Lens." *Journal for Research in Mathematics Education* 31, no. 4 (July): 454–82. https://doi.org/10.2307/749653.

National Council of Teachers of Mathematics (NCTM). 2014. *Principles to Actions: Ensuring Mathematical Success for All*. Reston, VA: NCTM.

Parks, Amy N. 2010. "Explicit Versus Implicit Questioning: Inviting All Children To Think Mathematically." *Teachers College Record* 112 (7): 1871–96. Retrieved from http://www.tcrecord.org/DefaultFiles/SendFileToPublic.asp?ft=pdf&FilePath=c:%5CWebSites%5Cwww_tcrecord_org_documents%5C38_15919.pdf&fid=38_15919&aid=2&RID=15919&pf=Content.asp?ContentID=15919.

Pólya, George. 1945. *How To Solve It: A New Aspect of Mathematical Method*. Princeton, NJ: Princeton University Press.

Schoenfeld, Alan H. 1992. "Learning To Think Mathematically: Sense-Making in Mathematics." In *Handbook for Research on Mathematics Teaching and Learning*, edited by Douglas A. Grouws, pp. 334–70. New York: MacMillan.

Selling, Sarah Kate. 2016. "Making Mathematical Practices Explicit in Urban Middle and High School Mathematics Classrooms." *Journal for Research in Mathematics Education* 47, no. 5 (November): 505–51.

Webb, David C. 2016. "Applying Principles for Active Learning to Promote Student Engagement in Undergraduate Calculus." *School of Education Faculty Contributions*. Retrieved from http://scholar.colorado.edu/educ_facpapers/4.

What Does Active Learning Mean for Mathematicians?

Benjamin Braun, Priscilla Bremser,
Art M. Duval, Elise Lockwood,
and Diana White

In August 2016, fifteen presidents of member societies of the Conference Board of the Mathematical Sciences (CBMS), an umbrella organization consisting of the American Mathematical Society and sixteen other professional societies in the mathematical sciences, released a statement on active learning [1] with the following call to action:

> We call on institutions of higher education, mathematics departments and the mathematics faculty, public policy-makers, and funding agencies to invest time and resources to ensure that effective active learning is incorporated into post-secondary mathematics classrooms.

This call is part of a broad movement to increase the use of active and student-centered teaching techniques across science, technology, engineering, and mathematics (STEM) disciplines. A landmark 2014 meta-analysis published in the *Proceedings of the National Academy of Sciences* [2] highlighted the efficacy of active learning techniques across STEM disciplines. In mathematics specifically, a comprehensive study of student outcomes for inquiry-based learning [3] has further established that active learning methods have a strong positive impact on women and members of other underrepresented groups in mathematics. This movement extends beyond the academic community—for example, at the federal level the White House STEM-for-All initiative [4] includes active learning as one of its three areas of emphasis for the 2017 budget.

While robust support from education researchers, funding agencies, public policy makers, and institutions is a critical component of effective active learning implementation, at the end of the day these techniques and methods are put into practice by mathematics faculty leading classes of students. Thus, mathematics faculty need to be well informed about active learning and related topics. Our goal in this article is to provide a foundation for productive discussions about the use of active learning in postsecondary mathematics. We focus on topics that frequently arise at the department level, namely, definitions of active learning, examples of active learning techniques and environments used by individual faculty or teams of faculty, things to expect when using active learning methods, and common concerns. An extended discussion of these issues and a substantial bibliography can be found in the six-part series on active learning [5] written by the authors for the AMS blog *On Teaching and Learning Mathematics*.

What Is Active Learning?

A frequently asked question is, what is active learning? We base our discussion on the definition given in the CBMS statement [1]: *Active learning [refers] to classroom practices that engage students in activities, such as reading, writing, discussion, or problem solving, that promote higher-order thinking.* Using a broad definition such as this increases the risk of faculty, administrators, and other stakeholders "talking past" one another, as much is left to the imagination regarding what actually happens with such methods. However, it also acknowledges that active learning can and does involve a wide variety of specific activities in diverse settings, with instructors of varied background and experience, and for different kinds of students.

Another approach to defining active learning is more useful in local settings, such as internal department discussions or conversations between department leaders and administrators. In this approach, one focuses discussion on a specific course and defines active learning as a task that students will complete during class time. This approach helps ensure that everyone in the discussion has a similar vision of what methods are actually being proposed or discussed in the context of explicit course goals and student-learning outcomes.

Examples of Active Learning Techniques and Environments

In contemporary college and university courses, lecturing remains the dominant teaching technique used by mathematics faculty. While active learning and lecture are sometimes viewed as two diametrically opposed teaching options, this is a misconception, as the following examples illustrate. We begin with examples that primarily involve individual faculty, and we end with examples that require collective buy-in and support from faculty, departments, and institutions.

THINK–PAIR–SHARE. One of the simplest examples of an active learning technique suitable for use in lectures is "think-pair-share." In this technique, the instructor provides students with a short task such as doing a computation, completing a step in a proof, generating one or more examples, or forming a hypothesis or conjecture. After providing the students with two to three minutes of time to independently consider the task ("think"), students take two minutes to compare their answers with other students sitting nearby ("pair"). Finally, some or all of the students are asked to share their answers in some manner, either with the groups next to them or with the entire class ("share"). Giving students time to think about and discuss mathematics midlecture encourages their active participation in the class. This task has no implications for departments or institutions and serves as an effective comprehension check in which students are able to refocus their attention during a lecture.

CLASSROOM RESPONSE SYSTEMS ("CLICKERS"). In addition to think-pair-share, there are many related examples of "classroom voting" systems and techniques that can be used to increase student engagement. These systems are often useful when scaling up think-pair-share and related techniques to large-lecture environments. While some systems are entirely Web- and mobile phone–based, others require students to rent or purchase a response device. Thus, depending on the choice of system, there can be implications for departments when clicker systems are widely used, and often it is helpful to implement clicker use with a team of faculty rather than individually.

INVERTED (OR "FLIPPED") CLASSES. In an inverted (or "flipped") classroom environment, instructor presentations of basic definitions, examples, proofs, and heuristics are provided to students in videos or in assigned readings that are completed before attending class. As a result, class time becomes available for active learning tasks that directly engage students. The type of task that instructors use during this time ranges from using think-pair-shares with complex problems or examples to having students work in small groups on a sequenced activity worksheet with occasional instructor or teaching assistant feedback. The inverted model of teaching has been used as the structure for entire courses, as an occasional event for handling topics that are less amenable to lecture presentations, as the basis for review sessions or problem-solving sessions, and more. Depending on the method used for flipping individual class periods or entire courses, department and/ or institutional support (in the form of technical assistance) may often be key ingredients in this model.

INQUIRY-BASED LEARNING. One of the most well-known active learning methods in mathematics is inquiry-based learning (IBL). In IBL courses, class time is spent with students working on problem sets individually or in groups, presenting solutions and/or proofs to the class, and receiving feedback from peers and faculty. IBL courses are not based on pure, unguided student discovery; instead, faculty design a series of carefully scaffolded (i.e., sequenced in a structured way) activities, some for individuals, some for pairs, some for small groups, and some for the whole class, including minilectures as appropriate. Because faculty using IBL need to develop facility with a range of teaching strategies and need to develop familiarity with many "teaching moves" that are not typically used in lecture environments, IBL is a more ambitious active learning environment. There are various opportunities for professional development with IBL, including the workshops offered by the Academy for Inquiry-Based Learning at www.inquirybasedlearning.org.

MATH EMPORIUM. The math emporium model uses a large room filled with computer workstations at which students progress through self-paced online courses. Unlike inverted classes, many emporium models do not include a lecture component at all, and most have been developed to handle remediation issues and low-level courses such as developmental mathematics and college algebra. An emporium usually has tables at which students can work collaboratively and is staffed by a large number of teaching assistants and tutors. Because the work of students is self-paced, students spend most of their time actively engaging with course content through a range of tasks. Because of the significant investment in classroom space and technological resources required, a math emporium is typically launched as a collaborative venture among faculty, departments, and administrators.

MODELING AND COMPUTER LABORATORIES. Modeling is a rich arena for increasing student engagement, one that is often augmented with computer labs. Since the 1990s, many mathematics courses have included computer lab activities for exploration using programs such as Mathematica, Maple, MATLAB, and Sage. Recent years have seen a growth in the number of support networks for faculty using lab and modeling components, such as the SIMIODE.org project for differential equations. The 2016 SIAM (Society for Industrial and Applied Mathematics) report *Guidelines for Assessment and Instruction in Mathematical Modeling Education* [8] provides examples of modeling activities across the undergraduate curriculum that actively engage students and discusses related issues such as assessment. Incorporating modeling and laboratory components into postsecondary courses can be done at many levels, ranging from stand-alone activities in a single class to program-wide implementations supported at the institutional level.

Things To Expect with Active Learning

Faculty using active learning for the first time need a realistic expectation of what impact these techniques will have. Because there are so many different active learning techniques and because different techniques often influence students in unique ways, it is not always possible to clearly say what will happen when we use a new active learning method. However, there do seem to be a few things faculty can typically expect. Here are five of them.

EXPECT TO GAIN INSIGHTS ABOUT YOUR STUDENTS. For many faculty using active learning, these techniques inspire richer discussions with students and provide a window into the reality of students' mathematical experiences. This richness allows faculty to be more responsive to students' misunderstandings, which in turn causes students to feel more supported in the course, frequently leading to increased engagement. Even in 200-student lectures, where student–faculty dialogue might be heavily moderated by clicker systems, faculty often report that active learning methods provide a clearer sense of what their students understand than with traditional lecture alone.

EXPECT YOUR STUDENTS TO SURPRISE YOU. Active learning provides opportunities for faculty–student interaction not present in courses focused on direct instruction. Active learning methods can reach and excite some students who might not typically be vocal or engaged in class—students who are quiet and reserved by nature frequently demonstrate their full potential when provided with the right opportunity. On the other hand, active learning methods can uncover deep misconceptions about mathematics, even from straight-A students, that homework and exams do not reveal. Furthermore, students often respond to active learning tasks with interesting observations and thought-provoking questions, infusing standard courses like calculus with fresh energy.

EXPECT RESISTANCE FROM SOME STUDENTS. For many reasons, it is common for some students to resist active learning methods, especially at the beginning of a course. Some students are not particularly interested in mathematics and do not want to engage at a deeper level. Other students have experienced significant success in traditional mathematics courses and feel threatened by an unfamiliar environment. With all students, instructors need to clearly articulate the value of the active learning methods they use and maintain high expectations for student participation and engagement. Often, students who are initially resistant find themselves surprised at the end of a course by how much they appreciate active learning.

EXPECT TO LEARN FROM YOUR MISTAKES. Much like learning mathematics, learning how to effectively use a new pedagogical technique, especially one of the more complex active learning techniques, involves a process of persistence and error correction through small failures. *Mathematics faculty need to be prepared to start small and develop gradually and consistently.* Almost every faculty member the authors have spoken with who uses active learning describes the development of his or her teaching as a sequence of mixed successes and failures. If you are implementing an active learning technique that is new to you, it is often helpful to first discuss with your department chair how teaching in that course will be evaluated for merit reviews. Many colleges have policies to support faculty as they build experience with new teaching techniques, especially if the techniques are evidence based.

EXPECT LONG-TERM IMPACTS. When used in combination with a foundation of good general teaching practices, active learning often has a particularly positive impact on student persistence and sense of belonging in mathematics. This in turn can lead students to be more engaged in their studies and pursue more mathematics over the long term. Because many active learning techniques emphasize communication and collaboration, faculty often report that using these techniques is a catalyst for building strong student communities. These peer networks persist through subsequent courses, contributing to students' experience throughout their mathematical studies. Many of these impacts of active learning become fully visible only after a course ends and thus can be hard to measure or even identify with standard course evaluation instruments.

Common Concerns about Active Learning

While active learning has many advocates among mathematicians, there are also responsible teachers who have reasonable concerns about active learning methods. We address four of them here.

HOW WILL STUDENTS LEARN THE MATHEMATICS IF WE DO NOT CLEARLY TELL THEM EVERYTHING ABOUT IT? The historical dominance of the lecture format rests on the belief that learning occurs as a result of transmission of information from instructor to student and that students learn by a process of taking in bits of information that their instructors say or write. Furthermore, because of our passion and love for mathematics, a natural human impulse is for mathematics faculty to tell students about the ways we have come to understand our discipline, to shed light on the subtleties that surround most mathematical ideas, and to explain the fundamental insights of our field. Our common experience, supported by research, demonstrates that learning is not this simple. For example, almost every teacher has experienced telling a student a certain mathematical fact—such as the fact that $(a + b)^2$ does not equal $a^2 + b^2$—only to have them demonstrate on a test that they have not learned it. Such experiences suggest that it is not enough for students simply to be told information if we want to produce deep and meaningful learning. Thus, the key is to find an effective balance between direct instruction and active learning, wherein instructors provide guidance through a combination of explanations and active learning tasks.

WHAT IF I CAN'T COVER THE SAME AMOUNT OF MATERIAL? Direct instruction alone can be an efficient way of getting through material. However, the example of students not knowing that $(a + b)^2$ does not equal $a^2 + b^2$ should not be far from our minds: lecturing in order to cover more material is not always effective for students. By exclusively considering course content coverage and responding to content coverage with telling, we risk forgetting the many other elements of student learning that active learning addresses, such as the cognitive goals for students outlined in the 2015 MAA *CUPM Curriculum Guide* [6], including:

- recognize and make mathematically rigorous arguments,
- communicate mathematical ideas clearly and coherently both verbally and in writing,
- work creatively and self-sufficiently,
- assess the correctness of solutions,
- create and explore examples,
- carry out mathematical experiments, and
- devise and test conjectures.

In addition to the recognition that content topics are not the exclusive subject of coverage, recent research suggests that coverage of material is less important for student persistence and achievement in mathematics than the use of teaching techniques that address these other types of learning goals.

HOW DO I KNOW IF I'M DOING A GOOD JOB WITH MY TEACHING? The crafting of rich lectures contributes to mathematicians' feelings of efficacy in their discipline. There are, however, a number of other ways in which teachers may gain efficacy while balancing traditional lecture with active learning in their classrooms. These methods include activities such as choosing problems, predicting student reasoning, generating and directing discussion, pushing students for high-quality explanations, asking for questions that extend student knowledge, and obtaining immediate feedback from students regarding what they just learned. Reflecting in this manner shifts the way we measure our own teaching away from the quality of our presentations and toward the quality of the tasks we provide students. Furthermore, many mathematicians who implement active learning report that they have a deeper understanding of student progress and can observe changes in students more clearly than in their previous courses.

I DIDN'T NEED ACTIVE LEARNING; WHY DO MY STUDENTS? Although this is changing, many mathematicians have not personally experienced undergraduate teaching environments that include active learning components. Thus, for many mathematicians and graduate students, their first experience with active learning techniques will be as teachers rather than as students. However, we should be careful when comparing our own experiences with those of our students; as Carl Lee [7] has written:

> I often engaged in math classes at a high cognitive level merely as a result of a teacher's direct instruction ("lecture"). As a teacher I quickly learned that I engaged few of my students by this process. Not all developed their "mathematical habits of mind" or "mathematical practices" through my in-class lectures and out-of-class homework (often worked on individually). I now better appreciate the significant role of personal context and informal education in the development of students' capacity.

Research [2], [3] suggests that active learning has a strong positive effect on a wide range of students, not only those who enter our courses ready to independently engage with math at a high cognitive level. That research also suggests that active learning does not harm, and may further benefit, already high-achieving students. Reflecting on our own educations, the authors agree that we would likely have built a firmer mathematical foundation had we experienced more active learning environments and that active learning would have prompted in us an earlier understanding of mathematics as an inquiry-based discipline.

Conclusion

New instructional techniques cannot be effectively implemented overnight. We must start small and develop gradually and consistently, ideally implementing changes as part of a team that can provide feedback and support. For experienced faculty, this development is something we need to do not only for ourselves but with an eye toward training the next generation of mathematicians.

Those of us who work at master's- and doctoral-granting institutions should provide graduate students with training and experience in using active learning techniques, whether as part of recitation duties or in situations where graduate students serve as independent instructors for courses. Given the many demands of graduate school, it is unreasonable to expect that every graduate student in math will emerge as an expert teacher, but we should provide as many opportunities as possible for graduate students to build their skill in using a combination of direct instruction and active learning techniques. For early-career faculty, long-standing professional development programs such as Project NExT of the Mathematical Association of America provide a valuable service to the mathematical community.

There is a fundamental way in which our training as mathematicians can help us develop as teachers: mathematicians are expert problem solvers. As a community of mathematicians and educators, we are in the process of solving the problem of how best to teach mathematics, and we are working together toward that end. As with all complex real-world problems, the challenge for us is that there is not an exact solution but rather a collection of approximate solutions. Nevertheless, our mathematical training has prepared us as problem solvers to hone our intelligence, our diligence, our spirit of curiosity, and our love of learning in order to develop meaningful and effective ways of teaching. These qualities are directly related to who we are as mathematicians, and they give us hope for success in our continued endeavor of improving mathematics teaching and learning for all.

References

[1] www.cbmsweb.org/Statements/Active_Learning_Statement.pdf (2016).

[2] S. Freeman, S. Eddy, M. McDonough, M. Smith, N. Okoroafor, H. Jordt, and M. Wenderoth, Active learning increases student performance in science, engineering, and mathematics, *Proc. Natl. Acad. Sci. USA* **111** (23), (2014), 8410–8415.

[3] M. Kogan and S. Laursen, Assessing long-term effects of inquiry-based learning: A case study from college mathematics. *Innov. High. Educ.* **39** (2014), 183–199.

[4] https://www.whitehouse.gov/blog/2016/02/11/stem-all. (2016).

[5] B. Braun, P. Bremser, A. Duval, E. Lockwood, and D. White, *Active Learning in Mathematics, Parts I–VI* [Web log posts]. Retrieved from blogs.ams.org/matheducation/tag /activelearning-series–2015 (2015).

[6] C. S. Schumacher and M. J. Siegel, co-chairs, and P. Zorn, editor, *2015 CUPM Curriculum Guide to Majors in the Mathematical Sciences*, MAA, Washington, DC, www.maa.org/programs /faculty-and-departments/curriculum-department-guidelines-recommendations/cupm (2015).

[7] C. Lee, *The Place of Mathematics and the Mathematics of Place* [Web log post]. Retrieved from blogs.ams.org/ matheducation/2014/10/01/the-place-of-mathematics-and-the -mathematics-of-place/ (2014).

[8] http://www.siam.org/reports/gaimme.php.

Written in Stone: The World's First Trigonometry Revealed in an Ancient Babylonian Tablet

Daniel Mansfield and N J Wildberger

The ancient Babylonians—who lived from about 4000 BCE in what is now Iraq—had a long forgotten understanding of right-angled triangles that was much simpler and more accurate than the conventional trigonometry we are taught in schools.

Our new research, published in *Historia Mathematica*, argues that the Babylonians were able to construct a trigonometric table using only the exact ratios of sides of a right-angled triangle. This is a completely different form of trigonometry that does not need the familiar modern concept of angles (Figure 1).

At school we are told that the shape of a right-angled triangle depends upon the other two angles. The angle is related to the circumference of a circle, which is divided into 360 parts or degrees. This angle is then used to describe the ratios of the sides of the right-angled triangle through sin, cos, and tan (Table 1).

But circles and right-angled triangles are very different, and the price of having simple values for the angle is borne by the ratios, which are difficult to compute and must be approximated.

This approach can be traced back to the Greek astronomer and mathematician Hipparchus of Nicaea (who died after 127 BCE). He is said to be the father of trigonometry because he used his table of chords to calculate orbits of the Moon and Sun.

But our new research shows that this was not the first, or only, or necessarily best approach to trigonometry.

FIGURE 1. The Plimpton 322 tablet. UNSW/Andrew Kelly, CC BY-SA. See also color image.

TABLE 1. The Three Ratios of a Modern Trigonometric Table, Rounded to Six Decimal Places, with Auxiliary Angle θ in Degrees (Daniel Mansfield, Author provided)

$\sin \theta$	$\cos \theta$	$\tan \theta$	θ
0.017452	0.999848	0.017455	1
0.034899	0.999391	0.034921	2
0.052336	0.998630	0.052408	3
0.069756	0.997564	0.069927	4
0.087156	0.996195	0.087489	5

Babylonian Trigonometry

The Babylonians discovered their own unique form of trigonometry during the Old Babylonian period (1900–1600 BCE), more than 1,500 years earlier than the Greek form.

Remarkably, their trigonometry contains none of the hallmarks of our modern trigonometry—it does not use angles, and it does not use approximation.

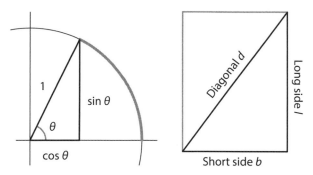

FIGURE 2. The Greek (left) and Babylonian (right) conceptualization of a right triangle. Notably, the Babylonians did not use angles to describe a right triangle. Daniel Mansfield, Author provided

The Babylonians had a completely different conceptualization of a right triangle. They saw it as half of a rectangle, and because of their sophisticated sexagesimal (base 60) number system, they were able to construct a wide variety of right triangles using only exact ratios (Figure 2).

The sexagesimal system is better suited for exact calculation. For example, if you divide one hour by three, then you get exactly 20 minutes. But if you divide one dollar by three, then you get 33 cents, with 1 cent left over. The fundamental difference is the convention to treat hours and dollars in different number systems: time is sexagesimal, and dollars are decimal.

The Babylonians knew that their sexagesimal number system allowed for many more exact divisions. For a more sophisticated example, 1 hour divided by 48 is 1 minute and 15 seconds.

This precise arithmetic of the Babylonians also influenced their geometry, which they preferred to be exact. They were able to generate a wide variety of right-angled triangles with exact ratios b/l and d/l, where b, l, and d are the short side, long side, and diagonal of a rectangle. The variety is a unique feature of their sexagesimal system and cannot be achieved using exact decimal ratios.

The ratio b/l was particularly important to the ancient Babylonians and Egyptians because they used this ratio to measure steepness.

The Plimpton 322 Tablet

We now know that the Babylonians studied trigonometry—which we take in a general sense to mean the measurement of triangles without the modern bias toward angles—because we have a fragment of one of their trigonometric tables.

Plimpton 322 is a broken clay tablet from the ancient city of Larsa, which was located near Tell as-Senkereh in modern-day Iraq. The tablet was written between 1822 and 1762 BCE.

In the 1920s, the archaeologist, academic, and adventurer Edgar J. Banks sold the tablet to the American publisher and philanthropist George Arthur Plimpton.

Plimpton bequeathed his entire collection of mathematical artifacts to Columbia University in 1936, and it resides there today in the Rare Book and Manuscript Library. It is available online through the Cuneiform Digital Library Initiative.

In 1945, the tablet was revealed to contain a highly sophisticated sequence of integer numbers that satisfy the Pythagorean equation $a^2 + b^2 = c^2$, known as Pythagorean triples.

This is the fundamental relationship of the three sides of a right-angled triangle, and this discovery proved that the Babylonians knew this relationship more than 1,000 years before the Greek mathematician Pythagoras was born (Figure 3).

Plimpton 322 has ruled space on the reverse side, which indicates that additional rows were intended. In 1964, the Yale-based science historian Derek J. de Solla Price discovered the pattern behind the complex sequence of Pythagorean triples, and we now know that it was originally intended to contain 38 rows in total (Figure 4).

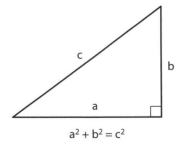

$$a^2 + b^2 = c^2$$

FIGURE 3. The fundamental relation between the side lengths of a right triangle. In modern times this is called Pythagoras' theorem, but it was known to the Babylonians more than 1,000 years before Pythagoras was born.

FIGURE 4. The other side of the Plimpton 322 tablet. UNSW/Andrew Kelly, Author provided. See also color image.

The tablet also has missing columns, and in 1981 the Swedish mathematics historian Jöran Friberg conjectured that the missing columns should be the ratios b/l and d/l (Table 2). But the tablet's purpose remained elusive.

The surviving fragment of Plimpton 322 starts with the Pythagorean triple 119, 120, 169. The next triple is 3,367, 3,456, 4,825. This makes sense when you realize that the first triple is almost a square (which is an extreme kind of rectangle), and the next is slightly flatter. In fact, the right-angled triangles are slowly but steadily getting flatter throughout the entire sequence.

So the trigonometric nature of this table is suggested by the information on the surviving fragment alone, but it is even more apparent from the reconstructed tablet.

This argument must be made carefully because modern notions such as angle were not present at the time Plimpton 322 was written. How then might it be a trigonometric table?

Fundamentally, a trigonometric table must describe three ratios of a right triangle. So we throw away sin and cos and instead start with the ratios b/l and d/l. The ratio that replaces tan would then be b/d or d/b, but neither can be expressed exactly in sexagesimal.

TABLE 2. The First Five Rows of Plimpton 322, with
Reconstructed Columns and Numbers Written in Decimal

		Information Relating to b/d or d/b			
b/l	d/l	$(d/l)^2$	b	d	Row
0.99166666	1.40833333	1.98340277	119	169	1
0.97424768	1.39612268	1.94915855	3,367	4,825	2
0.95854166	1.38520833	1.91880212	4,601	6,649	3
0.94140740	1.37340740	1.88624790	12,709	18,541	4
0.90277777	1.34722222	1.81500771	65	97	5

Instead, information about this ratio is split into three columns of exact numbers. A squared index and simplified values of b and d contain all the information about this third ratio that you might practically need, written without approximation.

No Approximation

The most remarkable aspect of Babylonian trigonometry is its precision. Babylonian trigonometry is exact, whereas we are accustomed to approximate trigonometry.

The Babylonian approach is also much simpler because it only uses exact ratios. There are no irrational numbers and no angles, and this means that there is also no sin, cos, tan, or approximation.

It is difficult to say why this approach to trigonometry has not survived. Perhaps it was the work of a lone genius, or perhaps this understanding was lost in 1762 BCE when Larsa was captured by Hammurabi of Babylon. Without evidence, we can only speculate.

We are only beginning to understand this ancient civilization, which is likely to hold many more secrets waiting to be discovered.

Quadrivium:
The Structure of Mathematics
as Described in Isidore
of Seville's Etymologies

ISABEL M. SERRANO, LUCY H. ODOM,
AND BOGDAN D. SUCEAVĂ

It is a paradox that, after the catastrophic CE fifth century, notorious for the great invasions of Western Europe and the Mediterranean Basin by populations coming from different geographical areas and the ultimate fall of the Roman Empire, the Roman culture continued to develop and several Latin works were produced. A natural question from the standpoint of the historian of mathematics is to understand how works written in this Roman postimperial historical period reflected the classical perspective on the structure of mathematics. Isidore of Seville was born in CE 560, approximately, and died on April 4, 636, in Visigothic Spain. His *Etymologies*, an encyclopedia compiled at the beginning of the seventh century, circulated widely in manuscript for many centuries and was printed as early as 1472. This encyclopedic work enjoyed a wide audience during the medieval period and became one of the most influential scholarly publications of its time. Because of its influence and accessibility in medieval libraries, it is interesting to examine the structure of mathematics, as described in *Etymologies*. In particular, we investigate the sources Isidore might have used when he wrote his etymological definitions.

Why Is Etymologies Important?

The Greco-Roman epoch witnessed important achievements in mathematics. Scholars investigated geometry and arithmetic to obtain subtle and interesting results through the classical Hellenistic and Roman periods, as most famously exemplified by Euclid's *Elements*. After the dissolution of the Western Roman Empire fostered by Gothic invasions, scholars located in the newfound cultural centers salvaged information, albeit less connected to the original scholarly context, through Latin translations.

The editors of the English edition of *Etymologies* [12] note in their *Introduction* (p. 24) that "it would be hard to overestimate the influence of *Etymologies* on medieval European culture, and impossible to describe it fully. Nearly a thousand manuscript copies survive, a truly huge number." They also point out that *Etymologies* "was among early printed books (1472), and nearly a dozen printings appeared before the year 1500." There are numerous papers and volumes where a thorough discussion of *Etymologies*'s influence is studied (for a series of references see pp. 30–31 in [12]). Of particular interest to mathematicians, however, is the encyclopedia's implicit description of mathematics as understood in Isidore's times. *Etymologies* aimed to summarize definitions known up until the early seventh century, a construction which, in philosopher Emil Cioran's suggestive wording, looks like a "a cemetery of definitions" [4]. Indeed, we might inquire today, if a phenomenon is defined, is the concept necessarily understood?

As historian Edward Grant remarks [9], "between the fourth and the eighth centuries, encyclopedic authors produced a series of Latin works that were to have significant influence throughout the Middle Ages, especially prior to 1200," among the most influential were Chalcidius, Macrobius, Martianus Capella, Boethius, Cassiodorus, Isidore of Seville, and the Venerable Bede. Thus, *Etymologies* was written during a peak of encyclopedic production. Despite the apparent success of encyclopedic work during this time, it is difficult to assess whether Isidore envisioned the impact his work would have over the course of the following centuries.

Among Latin encyclopedists, the core of scientific learning was referred to as the *quadrivium*, or the four mathematical sciences: arithmetic, geometry, astronomy, and music. The classical education model included grammar, rhetoric, logic, and the quadrivium to form the seven liberal arts. The quadrivium was initially outlined by Plato in *The Republic*, Book VII. In [12], p. 11, note 28, it is pointed out that "the scheme of the Seven Liberal Arts came to the Middle Ages primarily by way of Martianus Capella."

In [5] we read that: "Cassiodorus, Isidore, Augustine, and other Christian authors of antiquity and the Middle Ages, while warning of the dangers inherent in their pagan origins, encouraged or permitted study of the liberal arts as preparation for the understanding of Scripture and the study of theology." Hence, the education model inherited from the classical Greco-Roman period was adapted to educate potential clergymen, in Isidore's case in Visigothic Spain. Thus, we conclude that the *Etymologies* reflects theologians' views of mathematics in early Church history. Since *Etymologies* is a window into seventh-century mathematical thought, our fundamental inquiry is this: How was mathematics presented in the early medieval period?

Grant notes that in drafting *Etymologies*, Isidore "drew heavily upon Cassiodorus, who had, in turn, excerpted from the lengthy Boethian translation of Nicomachus's *Introduction to Arithmetic*" [9]. The editors of [12], supporting Grant's claim, stress the importance of Boethius' (480–524) works in Isidore's mathematical explanations ([12], p. 13). To this list of sources, we wish to add Boethius' lost Latin translation of Euclid's *Elements*. Because of the significant impact of Isidore's work, theological prestige, and his political influence, which later developed into centuries-long editorial success, we need to explore both Isidore's encyclopedic vision and the society in which he lived.

Culturally, Western Europe was experiencing drastic changes due to the fall of the Roman Empire. The Visigoths invaded the Roman Empire in 376 [1], defeating the Roman Imperial army several times [2] and sacking Rome under Alaric's leadership in 410. After 418, when the Visigoths settled in Aquitania, the Visigoths' political power extended south, covering Southern France and most of the Iberian Peninsula [11].

Isidore was related to the Visigoth royal family and is credited with persuading King Reccared I (586–601) to convert from Arianism (a faith named after Arius, c. CE 250–336, a Christian presbyter in Alexandria, Egypt) to Catholic Christianity. This was an important political step with profound historical implications. Likewise, Isidore participated in a Church council at Toledo in 610, and presided at the second Council of Seville, which took place in the year 618 or 619. Hence, Isidore evidently had a powerful political and religious presence in Visigothic Spain, and much of his editorial prestige is rooted in his political prestige.

Because of this historical and cultural context, the Visigoth ruling class adopted the Latin culture. Isidore, among other scholars, was admired for the extent of his knowledge of the Greek and Hebrew languages and developments ([12], p. 7). Yet, as Barney et al. note, Isidore's knowledge of the Greek language was limited ([12], p. 7). Thus, in transcribing his encyclopedia, Isidore relied on Latin translations of these sources.

The editors of [12] explain Isidore's process in writing the *Etymologies*:

> Obviously [Isidore] compiled the work on the basis of extensive notes he took while reading through the sources at his disposal. Not infrequently he repeats verbatim in different parts of the work; either he copied extracts twice or he had a filing system that allowed multiple use of a bit of information. Presumably he made his notes on the slips of parchment that he might have called *schedae*: ≪A *scheda* is a thing still being emended, and not yet redacted into books≫ (VI.xiv.8).

This description shows that Isidore was exposed to classical literature and relied heavily on these sources in constructing his work. Isidore's *Etymologies*, then, is not simply a summary, but also a commentary. The encyclopedia extends further by reflecting how information was preserved, evaluated, and incorporated into a more general image of the scholarly culture during the early seventh century. Very few works from this period remain, making *Etymologies* unique in its structure, content, and ability to reveal the past. This encyclopedia, in its hope of explaining terminologies to the intended audiences in Visigothic Spain, invites an analysis of the structure and arguments of knowledge during Isidore's time.

Furthermore, it is quite possible that at least for some parts of the book Isidore might have had help. The editors of [12] point out that "the guess that Isidore had help from a team of copyists finds some support in the fact that some errors of transmission may indicate that Isidore was using excerpts poorly copied or out of context, perhaps excerpts made by a collaborator" (see [12], p. 18, indicating also [8], p. 526).

Structure of Medieval Mathematics

In *Etymologies*, Book III is titled *De mathematica*, translated in [12] as *Mathematics*. Before engaging in the first etymological discussion, Isidore inserts a brief discussion of mathematics. He defines and classifies mathematics as follows [12], p. 89:

Mathematics in Latin means the science of learning (*doctrinalis scientia*), which contemplates abstract quantity. An abstract quantity is something that we investigate by reasoning alone, separating it by means of the intellect from matter or from accidental qualities such as even or odd, and other things of this kind. There are four types of mathematics, namely Arithmetic, Music, Geometry, and Astronomy. Arithmetic is the study of numeric quantity in and of itself. Music is the study that is occupied with the numbers that are found in sounds. Geometry is the study of size and shapes. Astronomy is the study that contemplates the course of the heavenly bodies and all the figures and positions of the stars. We will cover these studies, each in turn, a little more fully, that their principles can be suitabl[y] shown.

In the short paragraph III.ii, Isidore delves into a discussion on the "originators of mathematics," hinting at the scholars he referenced in compiling his work. He writes, "People say that Pythagoras was the first among the Greeks to commit the study of numbers to writing. Next, Nicomachus laid out the subject more broadly. Among the Latin speakers, first Apuleius and then Boethius translated this." Isidore indicates here among his references Nicomachus, followed chronologically by Apuleius and Boethius. Several authors note that Plato's *Timaeus* seems to have inspired certain parts in Isidore's text, particularly the section on Geometry, as indicated in [12], following J. Fontaine's study [8]. We add to these conclusions several connections to Euclid's *Elements*, which are not mentioned in [12]. Yet, any reader familiar with Euclid's work could easily notice that definitions in Isidore's text resembled, if not matched, those stated by Euclid. We gather that Isidore most likely used Boethius' translation of the *Elements*, which would have been completed a century earlier.

Since Boethius' translation of Euclid's *Elements* did not survive, it is not entirely clear if Boethius completed this translation. This uncertainty is primarily a result of Boethius' untimely execution in the middle of translation projects. Yet Isidore's access to Boethius' translation, despite the distance between the Visigothic (modern-day Spain) and the Ostrogothic (modern-day Italy) kingdoms, lies in their political, cultural, and religious ties, best exemplified through their shared Bible, translated into Gothic by Ulfila in the fourth century [3, 13, 14].

Due to the nature and the method of Isidore's *Etymologies*, it is quite possible that we are reading in Isidore's work fragments of Boethius' translation. We indicate where Euclid's definitions match Isidore's text in the following text.

ARITHMETIC

The first mathematical topic Isidore approaches is Arithmetic. The discussion begins with a purely etymological analysis of Latin and Greek numerals. Isidore mentions the classification of numbers into either evens or odds (he does not use "integers," but "numbers"), then he further discusses "evenly even" numbers, that is, multiples of 4, then "evenly odd" numbers, that is, numbers that in modern notation satisfy $4k + 2$, with k integer. The discussion is focused more on various classes of integers than on the study of their properties. Grant justly states [9], "Faced with an unrelated collection of inept definitions, supplemented by a few trivial examples, the reader of Isidore's section on arithmetic could have used little of it. A comparison with the arithmetic books of Euclid's *Elements* (Books VII to IX) illustrates the depth to which arithmetic had fallen."

Of particular interest, however, is III.v.11, where Isidore writes,

A perfect number is one that is completely filled up by its own parts, as, for example, 6, for it has 3 parts: 6, 3, and 2. The part that occurs 6 times is 1, the part that occurs 3 times is 2, and the part that occurs 2 times is 3. When these parts are added together, that is, when 1, 2, and 3 are summed up together, they make the number 6. Perfect numbers that occur within 10 include 6; within 100, 28; and within 1000, 496.

The definition of perfect numbers mimics Definition 22, Book VII, in Euclid's *Elements* ([7], p. 278), indicating some reference to concepts found in Euclid's work. (The editors of [12] do not point out this connection.) In a footnote on p. 90, they write, "In this book we often translate Isidore's term pars as ≪part,≫ where a modern equivalent would be ≪factor.≫ Isidore and his predecessors conceptualized multiplication not so much as a process that derives from factors, but rather as a set of static relationships among numbers." The notes to the edition ([7], p. 293), point out that "Theon of Smyrna and Nicomachus both give the same definition of a *perfect* number, as well as the law of formation of such numbers which Euclid proves in the later proposition IX.36."

We remark that Greek scholars were aware of the existence of a fourth perfect number, 8,128, which appears in the work of Nicomachus as early as CE 100, but Isidore did not list it. It is also worth observing that although Isidore maintains that there are three parts of 6, he initially cites the parts as 6, 3, and 2, which then change to 1, 2, and 3, which suggests a superficial understanding at best.

GEOMETRY

The structure of geometry is presented in III.xi, where Isidore divides geometry into four branches: planar, numerical magnitude, rational magnitude, and solid figures, which Grant describes as "strange" [9].

This division of geometry is indeed awkward and condensed. No remark is made on spherical geometry as a distinct discipline (not even in the section on astronomy) and explanations of geometric proofs and their structure (i.e., the relationship between geometry and logic) are altogether absent, although the classical Greco-Roman period had advanced knowledge in these directions.

In III.xi.2, Isidore writes, "Planar figures are those that have length and breadth and are, following Plato, five in number." The editors of [12] point out that in *Timaeus* Plato presents five solids, and not five planar figures. What Isidore intended is ambiguous because he does not specify five planar figures. Isidore continues with the specification that "numeric size is that which can be divided by the numbers of arithmetic." Then, the single sentence present in III.xi.3 introduces a distinction for one of the categories listed in III.xi.1: "rational sizes are those whose measures we are able to know, but irrational sizes are those the quantity of whose measures cannot be known."

Paragraph III.xii includes definitions of geometrical figures, with the solids included in the same paragraph as the planar figures. The figures are introduced in the following order: the *cube* (as a particular case of solid figure), the *circle* (lacking a clear definition of any geometric consistency), then the *quadrilateral*, a *planar figure* in general, described by the term *dianatheton grammon*, then an *orthogonium*, which is "a planar figure with a right angle," which is also a triangle, which is in modern language a right triangle. The section continues by mentioning the *isopleuros*, that is, a planar figure, that could be either an isosceles or an equilateral triangle, a *sphere*, a *cube* again, a *cylinder*, a *cone*, and a *pyramid*.

From this list we see that the classification is not complete, and no indication of a method in selecting this information is suggested.

Grant [9] points out that the definition of the cube was given by Euclid as a "solid figure contained by six equal squares," which is quite precise, whereas Isidore defines a cube as "a proper solid figure which is contained by length, breadth and thickness," a definition satisfied by every solid. On the other hand, Isidore defines a quadrilateral as "a square in a plane which consists of four straight lines." Grant writes (see [9]) that this equates "all four-sided figures with squares!" Rather unusual is the definition of a cylinder: "a four-sided figure having a semicircle above." Isidore (and his collaborators) collected the definitions, but, as is apparent, lacked experience in using these concepts.

In the last part of III.xii, we observe clear references to Euclid's work. The last paragraph starts by saying that the first figure of the art of geometry is the point (*punctus*), which has no parts. He continues: "The second figure is the line (*linea*), a length without breadth." We recognize in these statements the first two definitions in Book I of Euclid's *Elements*. However, Isidore next states: "A straight (*rectus*) line is one that lies evenly along its points." There is little doubt this is exactly Definition 4, Book I, in Euclid's *Elements* ([6], p. 165). We note that here the copyist skipped Euclid's Definition 3, "The ends of a line are points" ([6], p. 165). This is particularly important, as it shows how in Euclid lines (corresponding to modern curves) are finite, as long as they have ends. For our present discussion, the omissions are as important as Isidore's selections.

The text continues as follows: "A plane (*superficies*) has length and breadth only." We believe that the editors of [12] should have translated *superficies* as "surface," as this is the term used in Definition 5 of Euclid's *Elements*. In this definition, Euclid described a more general surface, not only a plane (see also the important comments from [6], p. 169). For the plane, there is a separate definition (it has number 7 in the *Elements* [6], p. 171). The very next sentence in Isidore's *Etymologies* is, "The boundaries of planes (*superficies*) are lines." This matches Definition 6 in Euclid's Elements, suggesting again that the English translation should be "surface" instead of "plane."

Then Isidore stops abruptly. He writes one more sentence, of a rather obscure intention: "The forms of these are not placed in the preceding ten figures, because they are found among them." The editors of [12] usefully footnote the following: "That is, the point, line, and plane are illustrated by the manuscript figures of the circle, the various planes, and the solids, respectively."

Isidore's list of concepts retained in section III.xii is not exhaustive, in that the copyist would have come across more of Euclid's definitions, which are not discussed, for example, definitions for the concepts of angle, diameter, rhomboid, parallel lines, etc. We find it remarkable that many definitions available in Book I of the *Elements* are not included in *Etymologies*. However, it is difficult to draw any conclusions without knowing the contents of the translation Isidore used. It is possible that Boethius translated just Book I, or just parts of it. If Boethius completed a translation of the entire *Elements*, then Isidore chose the concepts he would include.

If Isidore had access to a complete version of the *Elements*, then his omissions demonstrate (i) Isidore's lack of interest in expanding the geometric section, (ii) Isidore's unfamiliarity with Euclid's work, and (iii) as *Etymologies* served as the educational model for Visigothic clergy, that the clergy needed in Isidore's vision a basic understanding of only the aforementioned geometric concepts.

Music

Isidore follows a classical model first introduced by Plato, who viewed music as a branch of mathematics. There are several studies addressing "the place of Musica in medieval classifications of knowledge," for example [5], where we read that "models for the medieval classifications were available in Greek philosophy, Latin authors of Late Antiquity (Boethius), and in the writings of Arabic philosophers and commentators on Aristotle."

Isidore notes that "music is the practical knowledge of modulations and consists of sound and song." The editors of [12] comment on III. xv that "several early manuscripts present elaborate figures, of obscure meaning, illustrating various mathematical principles of music. [. . .] Presumably because his work is incomplete, Isidore does not discuss these in his text."

After Isidore covers three divisions of music (harmonic, *organicus*, and rhythmic), he reaches the topic of musical numbers, which was to act as the primary connection of music to mathematics. For this reason, we cite here the whole paragraph III.xxiii ([12], pp. 98–99):

Musical Numbers (*De Numeris Musicis*)

1. You find numbers with respect to music in this way (see viii.3 above). When the high and the low numbers have been set, as, for example, 6 and 12, you see by how many units 6 is exceeded by 12, and that is by 6 units. You make this number into a square, and 6 six times make 36. You add together the low and high numbers that you first took, 6 and 12, and together they make 18. You divide 36 by 18, and it makes 2. You add this to the low number, that is, 6, and it comes to 8. Eight is the mean between 6 and 12. Wherefore 8 exceeds 6 by two units, that is, a third of 6, and 8 is exceeded by 12 by four units, a third of 12. Thus, the high number exceeds the mean by the same proportion as the low number is exceeded by the mean.

2. But just as this proportion in the universe derives from the revolution of the spheres, so even in the microcosm it has such power beyond mere voice that no-one exists without its perfection and lacking harmony. Indeed, by the perfection of the same art of Music, meters are composed of arsis and thesis, that is, by rising up and setting down.

We can only conjecture that this section is incomplete. Some of the assertions included here are a discussion of particular cases without a clear context. If Isidore inherited the view that music is a branch of mathematics from earlier references, his explanation lacks a convincing argument, as demonstrated in the earlier discussion.

Astronomy

We do not intend to engage here in a discussion on the Ptolemaic model of the universe. Instead, we mention only that a geocentric model of the universe was considered in the *Etymologies* as the fourth branch of mathematics.

In III.xxiv, Isidore states that "astronomy is the law (cf. νόμος, "law") of the stars (*aster*,) which, by investigative reasoning, touches of the courses of the constellations, and the figures and positions of the stars relative to each other and to the earth." Then Isidore covers a description of the universe following Claudius Ptolemy's model. The *Almagest* was barely known during the early Middle Ages, and the fairly popular book by Johannes de Sacrobosco (his *Tractatus de Sphaera*) appeared much later, in the thirteenth century, and was far more elementary. The editors of [12] point out (see p. 99) that Isidore identifies Claudius Ptolemy (CE second century) with the Ptolemys who ruled Egypt (and this is not the only historical information that lacks accuracy). The description of Ptolemy's system is of less mathematical interest today; that is true for reasons illustrated by statements such as III.lxi: "Stars are said not to possess their own light, but to be illuminated by the sun, as the moon is."

We can only agree with Grant when he writes in [9], p. 17, that "too often, [Latin encyclopedists] failed to comprehend the material they read; nonetheless, they copied it, or paraphrased it, in their own treatises." Even the structure of mathematics has been inherited from classical Greek literature, without much evolution, reflecting a halt in academic progress. However, we must note that the positive contribution of Latin encyclopedic works rests in their collection and preservation of Greco-Roman knowledge that otherwise would have been lost in Western Europe. Unfortunately, the lack of translated classical scientific works affected the understanding of fundamental concepts.

Isidore's official biographies characterize him as one of the most influential scholars of the early centuries. *Etymologies*, like its author, had a long-lasting effect. The literary work survived in ancient libraries for centuries and served as a model of reference and structural organization of knowledge for many centuries. Only five centuries after Isidore's death did Western Europe witness new attempts of original scientific inquiries [9, 10]. Even then, these groundbreaking scholars were educated in places where *Etymologies* was preserved and held in high regard. Thus, *Etymologies*, in all its influence, demonstrates and illustrates the ample regression in knowledge in Latin Europe during the centuries following the fall of the Roman Empire. For the mathematician investigating the complex historical evolutions of understanding mathematics across the centuries, the content of the *Etymologies* is revealing and symptomatic. A collection of information does not and cannot replace the competent active use of mathematics. This process can be partially explained by the historical decay happening all over Europe after the fall of the Roman Empire, but only by looking at the specific pieces of mathematical information can we assess the precise measure of this extensive loss of competences and profound mathematical thinking.

References

[1] Ammianus Marcellinus, *The Later Roman Empire* (CE 354–378), selected and translated by Walter Hamilton, introduction and notes by Andrew Wallace-Hadrill, Penguin Books, London, second reprint, 2004.

[2] A. Barbero, *The Day of the Barbarians. The battle that led to the fall of the Roman empire*, translated by John Cullen, Walker & Company, New York, 2007.

[3] J. L. Borges, with the help of María Esther Vazquez, *Literaturas germánicas medievales*, 1966, Falbo, Buenos Aires, Argentina.

[4] E. M. Cioran, *Précis de décomposition (A Short History of Decay)*, Ed. Gallimard, Paris, 1949.

[5] J. Dyer, *The Place of Musica in Medieval Classifications of Knowledge*, The Journal of Musicology, Vol. 24, No. 1 (Winter 2007), pp. 3–71.

[6] Euclid, *The Thirteen Books of Euclid's Elements, translated from the text of Heiberg, with introduction and commentary by Sir Thomas L. Heath*, Second Edition, Vol. I, Dover Publ. Inc., New York, 1956.

[7] *The Thirteen Books of Euclid's Elements, translated from the text of Heiberg, with introduction and commentary by Sir Thomas L. Heath*, Second Edition, Vol. II, Dover Publ. Inc., New York, 1956.

[8] J. Fontaine, *Isidore de Seville et la mutation de l'encyclopédisme antique*, in Cahiers d'historie mondiale IX, Neuchâtel, pp. 519–538, reprinted in the volume *Tradition et actualité chez Isidore de Séville*, Variorum, London, 1988.

[9] E. Grant, *The Foundations of Modern Science in the Middle Ages: Their Religious, Institutional and Intellectual Contexts* (Cambridge Studies in the History of Science), Cambridge University Press, Cambridge, U.K., 1996.

[10] E. Grant, *Science and Religion, 400 B.C. to A.D. 1550: From Aristotle to Copernicus*, Johns Hopkins University Press, Baltimore, 2006.

[11] Isidore of Seville, *History of the Goths, Vandals, and Suevi*, translated by Guido Donini and Gordon B. Rod, E. J. Brill, Leiden, Netherlands, 1970.

[12] Isidore of Seville, *The Etymologies*, Translated, with Introduction and Notes, by Stephen A. Barney, W. J. Lewis, J. A. Beach, and Oliver Berghof, Cambridge University Press, Cambridge, U.K., 2006.

[13] Pierre Riché, *Éducation et culture dans l'Occident barbare, VIe-VIIIe siècle*, quatrième édition, revue et corrigée, Éditions du Seuil, Paris, 1995.

[14] E. A. Thompson, *The Visigoths in the time of Ulfila*, second edition, with a foreword by Michael Kulikovski, Gerard Duckworth & Co. Ltd., London, 2008.

The World War II Origins of Mathematics Awareness

Michael J. Barany

Since ancient times, advocates for mathematics have argued that their subject is foundational for many areas of human endeavor, though the areas and arguments have changed over the years. Much newer, however, is the idea that mathematicians should systematically try to promote the usefulness or importance of mathematics to the public. This effort, which I shall generically call "mathematics awareness," was largely an American invention. One outward manifestation was the 1986 inauguration, by President Ronald Reagan, of the first Mathematics Awareness Week. Every year since then, mathematicians and mathematics educators in the United States have dedicated a week—or, beginning in 1999, the month of April—to raising public awareness of "the importance of this basic branch of science to our daily lives," as Reagan put it.

While today's mathematics awareness is focused on schools and on peaceful applications of mathematics, a direct line connects it to its origins in a very different kind of activity: mathematicians promoting their expertise to leaders of the American war effort during World War II. Recent mathematics awareness has focused on encouraging more people to take up the discipline. However, wartime and early postwar mathematics awareness centered on securing resources for those already in the profession. To understand the origins of mathematics awareness, one must follow the money.

A Discipline in Need

For most of the discipline's history, mathematicians have supported their research either through independent wealth or through patronage from the wealthy. Universities and a select few other academic

A WAR OF MATHEMATICS

World War II was not the first war that mathematicians attempted to characterize as "a war of mathematics," but it was the first one where the characterization appeared to stick. In addition to lobbying elite policy makers, mathematicians wrote articles for the popular press and offered radio broadcasts that attempted to explain why mathematics mattered in terms the masses could understand. For example, Bennington P. Gill, who served as AMS treasurer from 1938 to 1948, gave an interview in 1942 with WNYC for their series on *The Role of Science in War* (www.wnyc.org/story/bennington-p-gill/).

institutions—all, themselves, historically channels for wealthy patronage—eventually became the dominant sites and funders of mathematical scholarship. So long as publication and travel were relatively small parts of such scholarship, this arrangement suited mathematicians' needs well enough. But by the early twentieth century, mathematicians were publishing and traveling much more than before and across greater distances. They needed new organizations and new sponsors to support their work.

Such were the rationales for mathematicians' first professional societies, many of which date to the latter part of the nineteenth century and the start of the twentieth. The American Mathematical Society (AMS) originated in 1888 on the heels of corresponding societies in Europe, such as the London Mathematical Society (1865) and Société Mathématique de France (1872). The Mathematical Association of America (MAA) entered the scene at the close of 1915. These societies drew their support principally from their members' universities (both directly and by way of members' dues), national governments, and private sponsorship.

Following the First World War, U.S. mathematicians had some limited success securing corporate sponsorship for their work, for instance from the American Telephone & Telegraph Company, and considerably greater success courting major philanthropies such as the Rockefeller Foundation and the Carnegie Corporation of New York. Yet these relationships tended to be piecemeal and tenuous. U.S. mathematicians had successfully bid to host the 1924 International Congress of Mathematicians but ended up ceding the congress to Toronto after finding themselves unable to secure the needed financial backing. (John Charles Fields, on the other hand, managed to find enough money for the Toronto meeting that it concluded with a modest surplus, which provided the seed money for what became the Fields Medals.)

The Americans tried again to host an International Congress in 1940. When preliminary fundraising efforts again fell short, Institute for Advanced Study mathematician Marston Morse approached the Rockefeller Foundation with the argument that his discipline was "unique . . . as having no natural sources of support." Indeed, when it came to major donors, Rockefeller and Carnegie were the only relatively sure bets, and no grant was assured. To grow, U.S. mathematics would need new constituencies and new sources of funding.

Preparing for War

The AMS suspended plans for the envisioned 1940 International Congress of Mathematicians following the German invasion of Poland in 1939. As war threatened to engulf Europe and beyond, U.S. mathematicians thought back to their experiences of the Great War. Some concluded that a lack of coordination among U.S. mathematicians had restricted their contributions to the previous war effort. Without such coordination, military leaders would have a hard time learning where mathematicians were needed and where the needed mathematicians could be found.

A new joint AMS–MAA War Preparedness Committee aimed to provide this coordination by synthesizing the lessons from the last war and positioning mathematicians for a new conflict that seemed sure to draw American involvement sooner or later. Marston Morse was appointed chair, and a subcommittee chaired by Dunham Jackson focused on mathematical research. The subcommittee included Marshall Stone, whose two-year presidency of the AMS would fall in the middle of the United States' official engagement in World War II. From start to finish, Stone advocated formal mathematical coordination with particular force and frequency. In a summer 1940 missive on "the organizational aspects of the research problems of national defense," Stone articulated three purposes for the subcommittee. First and most urgent was to find an efficient means to join together current "technical problems and competent mathematicians" who could solve them. Second, the United States would need to make much greater use of mathematical techniques than it currently did. The third, long-range, goal was to make war service pay off for the U.S. mathematical profession even after the war's end.

"If mathematics is to be brought to bear upon our defense problems in full measure," Stone then asserted, "we shall have to organize and conduct propaganda to this end." He anticipated an uphill struggle. The subcommittee would have to confront "not only the appalling limitations of our military officers, but also the general American attitude of antagonism to theory in general and to mathematical refinements in particular and the abysmal ignorance of the majority of intelligent Americans concerning the uses of mathematics."

Enter Mina Rees

The U.S. military officially entered the war as 1941 drew to a close. The next year, the AMS and MAA responded by dissolving the War Preparedness Committee and appointing a new War Policy Committee, with Marshall Stone (soon to be AMS president) as chair and Marston Morse (who was just finishing his own term as AMS president) in a supporting role. Soon, leaders from academia, philanthropy, and the military drew on approaches from their respective fields to develop for the U.S. government a system for identifying problems and contracting them out to academic research groups. This formed the basis for a massive system of contracts that would support advanced mathematical research and training after the war, as well as the postwar system of government grants familiar to many mathematicians today. But mathematics awareness remained the exclusive province of a narrow elite.

Richard Courant, one of the academic leaders who helped to craft that system, used his wartime government connections both during and after the war to build his institute at New York University into one of the world's leading centers of mathematics. Perhaps even more important for postwar mathematics in and beyond the United States, however, was Courant's close associate Mina Rees. Although she had earned a Ph.D. under Leonard Dickson from the University of

A Few Key Men at the Top

Marshall Stone's dim view of the public appreciation for mathematics led him to focus on a few "key men at the top" rather than aim to convince the masses or even the much smaller mass of officers and policy makers. Among those key men were Harvard president James Bryant Conant, chair of the National Defense Research Committee, and Frank Jewett, chairman of the Board of Directors of Bell Laboratories and president of the National Academy of Sciences. Referring to the commonplace characterization of the Great War as the chemist's war, Conant famously quipped on the front page of *Chemical & Engineering News* in November 1941, "This is a physicist's war rather than a chemist's." According to AMS secretary Roland Richardson, when Conant shared the view with Jewett, the latter shot back that "It may be a war of physics, but the physicists say it is a war of mathematics." At least one key man got the message.

Further Reading: Michael J. Barany, "Remunerative combinatorics: Mathematicians and their sponsors in the mid-twentieth century," in *Mathematical Cultures: The London Meetings 2012–2014*, edited by Brendan Larvor (Basel, Switzerland: Birkhäuser, 2016, pp. 329–346; preprint online at mbarany.com/publications.html).

Chicago, Rees's prospects within the mathematics profession were limited by widespread institutional sexism. At Courant's urging, Rees was appointed as technical aide to the main government clearinghouse for coordinating mathematicians' war service. There, she facilitated the broad array of contracts by gathering information, assessing outcomes, and making needed connections.

Stone, in 1944, expressed his frustration that the government "would display considerable reluctance to call on the leaders of our profession." In his view, mathematicians could and should have done much more to dedicate themselves wholly to the war effort. He himself set out immediately after the conclusion of his term as AMS president on a mission classified top secret to advise and assess Allied signal intelligence in India, Burma, and China in the first part of 1945.

Mathematicians in the United States concluded the war with a range of views of their relative success or failure. While they did not lay claim to a breakthrough on the scale of the Manhattan Project as physicists would, mathematicians contributed to a great many of the United States' decisive wartime innovations in weaponry, aeronautics, provisioning, communications and intelligence, and other areas.

Coordination for a Growing Discipline

In 1946, Rees took over the mathematics arm of the Office of Naval Research (ONR), helping the navy to become a leading funder of research and publication in both pure and applied mathematics. One of her earliest efforts was to forge a partnership with the AMS's *Mathematical Reviews* to translate new Russian mathematical works into English. This and related undertakings reinforced the United States as an international clearinghouse for postwar mathematics, allowing U.S. institutions to assume a lasting dominance in the discipline.

The ONR led the way for a wide range of government-funded research programs associated with other branches of the military and various civilian offices. These eventually included the National Science Foundation (NSF), founded in 1950 after years of debate informed by wartime and early postwar military-sponsored science. ONR and other contracts funded faculty, students, seminars, and publications on a wide scale in the first postwar decades. They supported visiting researchers from abroad and allowed U.S. mathematicians to disseminate their work rapidly and efficiently. These new funding sources allowed the mathematics profession to grow quickly—so quickly, in fact, that U.S. mathematicians could turn from worrying about finding enough sources of support to worrying about finding enough people to make good use of that support.

Over the years, as new mathematical organizations sprang up and the mathematical profession collectively faced new funding and policy issues, the AMS and MAA continued to adapt and reconfigure a series of joint undertakings aimed at the kind of mathematics awareness pursued in the 1940s.[1] Early postwar joint committees laid the groundwork for the International Mathematical Union and advocated for the new NSF and a mathematics division in the National Research Council. The 1970s-era Joint Projects Committee in Mathematics eventually

MINA REES

In recognition of Rees's contributions to the mathematics profession, the AMS Council adopted a resolution in 1953 asserting:

> Under her guidance, basic research in general, and especially in mathematics, received the most intelligent and whole-hearted support. No greater wisdom and foresight could have been displayed and the whole postwar development of mathematical research in the United States owes an immeasurable debt to the pioneering work of the Office of Naval Research and to the alert, vigorous and farsighted policy conducted by Miss Rees. (*Bulletin of the AMS*, March 1954.)

Rees continued after 1953 to serve as an administrator and advisor for a wide range of important boards and institutions, fostering the development of pure and applied mathematics in and beyond the United States.

Further reading: Judy Green, Jeanne LaDuke, Saunders Mac-Lane, and Uta C. Merzbach, "Mina Spiegel Rees (1902–1997)," *Notices of the American Mathematical Society* 45 (1998) no. 7, 866–873; Amy Shell-Gellasch, *In Service to Mathematics: The Life and Work of Mina Rees* (Boston: Docent Press, 2011).

grew into the Joint Policy Board for Mathematics, which currently spearheads the annual Mathematics Awareness Month (in 2017, the name became Mathematics and Statistics Awareness Month).

A New Wave

If more people saw how important math was to their lives, the thinking since the 1980s has been, more people would participate in shaping that mathematics in the future. In practice, it has not always worked out that way. National and even international campaigns for mathematics education have put the power of mathematics on public display, but for many there remain significant barriers to success in the mathematical profession.

The first wave of mathematics awareness took place at a time when U.S. mathematicians were relatively isolated from many areas of policy and the economy and correspondingly lacked obvious places to turn for resources. The second wave responded to a different problem: the need for a public posture that would ensure a supply of mathematicians in the future. The Mathematics Awareness Weeks, and later Months, addressed that need by displaying the relevance of mathematics to modern life and to society at large.

We are now starting to see a third wave of mathematics awareness, focused less on what mathematics can do for war-making policy elites or for everyday citizens and more on the mathematicians themselves who create and apply that mathematics and train future generations in their field. Efforts to highlight women and minority pioneers in the discipline and to support underrepresented groups may yet make headway where other approaches have fallen short. The history of mathematics awareness shows that by banding together and working systematically, mathematics organizations can make real changes in how important constituencies view and engage the discipline. The history also shows that mathematics awareness can take many forms, each reflecting the priorities and blind spots of its time.

Note

1. For a blow-by-blow account of the various joint committees, see Everett Pitcher, A History of the Second Fifty Years: American Mathematical Society, 1939–1988 (Providence, RI: American Mathematical Society, 1988), pp. 273–85.

FIGURE 1 from "How to Play Mathematics" (Wertheim).
Photo © Institute for Figuring.

FIGURE 2 from "How to Play Mathematics" (Wertheim). Courtesy Wikipedia.

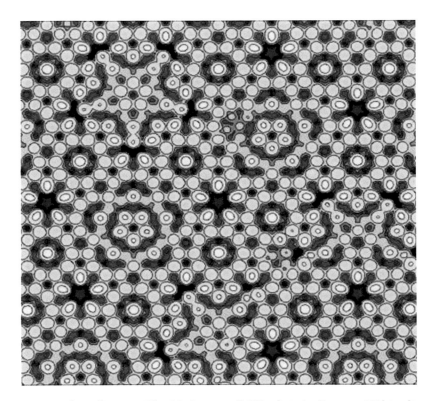

FIGURE 4 from "How to Play Mathematics" (Wertheim). Courtesy Wikipedia.

Encoding

$(x_1 \vee x_2 \vee x_3) \wedge (\bar{x}_1 \vee \bar{x}_2 \vee \bar{x}_3) \wedge (x_1 \vee x_3 \vee x_4) \wedge (\bar{x}_1 \vee \bar{x}_3 \vee \bar{x}_4) \wedge$
$(x_1 \vee x_4 \vee x_5) \wedge (\bar{x}_1 \vee \bar{x}_4 \vee \bar{x}_5) \wedge (x_2 \vee x_3 \vee x_5) \wedge (\bar{x}_2 \vee \bar{x}_3 \vee \bar{x}_5) \wedge$
$(x_1 \vee x_5 \vee x_6) \wedge (\bar{x}_1 \vee \bar{x}_5 \vee \bar{x}_6) \wedge (x_2 \vee x_4 \vee x_6) \wedge (\bar{x}_2 \vee \bar{x}_4 \vee \bar{x}_6) \wedge$
$(x_1 \vee x_6 \vee x_7) \wedge (\bar{x}_1 \vee \bar{x}_6 \vee \bar{x}_7) \wedge (x_2 \vee x_5 \vee x_7) \wedge (\bar{x}_2 \vee \bar{x}_5 \vee \bar{x}_7) \wedge$
$(x_3 \vee x_4 \vee x_7) \wedge (\bar{x}_3 \vee \bar{x}_4 \vee \bar{x}_7) \wedge (x_1 \vee x_7 \vee x_8) \wedge (\bar{x}_1 \vee \bar{x}_7 \vee \bar{x}_8) \wedge$
$(x_2 \vee x_6 \vee x_8) \wedge (\bar{x}_2 \vee \bar{x}_6 \vee \bar{x}_8) \wedge (x_3 \vee x_5 \vee x_8) \wedge (\bar{x}_3 \vee \bar{x}_5 \vee \bar{x}_8) \wedge$
$(x_1 \vee x_8 \vee x_9) \wedge (\bar{x}_1 \vee \bar{x}_8 \vee \bar{x}_9) \wedge (x_2 \vee x_7 \vee x_9) \wedge (\bar{x}_2 \vee \bar{x}_7 \vee \bar{x}_9) \wedge$
$(x_3 \vee x_6 \vee x_9) \wedge (\bar{x}_3 \vee \bar{x}_6 \vee \bar{x}_9) \wedge (x_4 \vee x_5 \vee x_9) \wedge (\bar{x}_4 \vee \bar{x}_5 \vee \bar{x}_9)$

Case split as binary tree

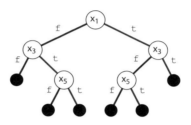

Figure 1 from "The Science of Brute Force" (Heule/Kullmann).

Proof | Unit clause justification

Proof	Unit clause justification
$(x_1 \vee x_3)$	$(x_1 \vee x_2 \vee x_3)$, $(x_1 \vee x_3 \vee x_4)$, $(\bar{x}_2 \vee \bar{x}_4 \vee \bar{x}_6)$, $(x_1 \vee x_6 \vee x_7)$, $(x_3 \vee x_6 \vee x_9)$, $(\bar{x}_2 \vee \bar{x}_7 \vee \bar{x}_9)$
$(x_1 \vee x_5)$	$(x_1 \vee x_3)$, $(x_1 \vee x_4 \vee x_5)$, $(x_1 \vee x_5 \vee x_6)$, $(\bar{x}_2 \vee \bar{x}_4 \vee \bar{x}_6)$, $(x_2 \vee x_5 \vee x_7)$, $(\bar{x}_3 \vee \bar{x}_4 \vee \bar{x}_7)$
(x_1)	$(x_1 \vee x_3)$, $(x_1 \vee x_5)$, $(\bar{x}_2 \vee \bar{x}_3 \vee \bar{x}_5)$, $(\bar{x}_3 \vee \bar{x}_5 \vee \bar{x}_8)$, $(x_2 \vee x_6 \vee x_8)$, $(x_1 \vee x_8 \vee x_9)$, $(\bar{x}_3 \vee \bar{x}_6 \vee \bar{x}_9)$
$d(x_1 \vee x_3)$	
$d(x_1 \vee x_5)$	
$(\bar{x}_3)$	(x_1), $(\bar{x}_1 \vee \bar{x}_2 \vee \bar{x}_3)$, $(\bar{x}_1 \vee \bar{x}_3 \vee \bar{x}_4)$, $(x_2 \vee x_4 \vee x_6)$, $(\bar{x}_1 \vee \bar{x}_6 \vee \bar{x}_7)$, $(\bar{x}_3 \vee \bar{x}_6 \vee x_9)$, $(x_2 \vee x_7 \vee x_9)$
$(\bar{x}_5)$	(x_1), $(\bar{x}_3)$, $(\bar{x}_1 \vee \bar{x}_4 \vee \bar{x}_5)$, $(\bar{x}_1 \vee \bar{x}_5 \vee \bar{x}_6)$, $(x_2 \vee x_4 \vee x_6)$, $(\bar{x}_2 \vee \bar{x}_5 \vee \bar{x}_7)$, $(x_3 \vee x_4 \vee x_7)$
$\perp$	(x_1), $(\bar{x}_3)$, $(\bar{x}_5)$, $(x_2 \vee x_3 \vee x_5)$, $(x_3 \vee x_5 \vee x_8)$, $(\bar{x}_2 \vee \bar{x}_6 \vee \bar{x}_8)$, $(\bar{x}_1 \vee \bar{x}_8 \vee x_9)$, $(x_3 \vee x_6 \vee x_9)$

Figure 2 from "The Science of Brute Force" (Heule/Kullmann).

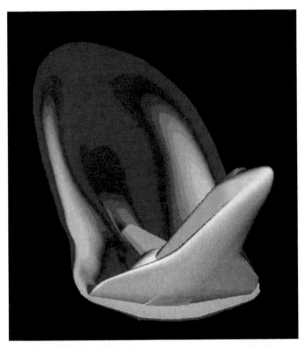

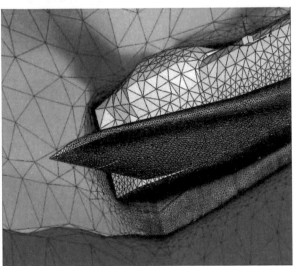

Figure 1 from "Computational Thinking in Science" (Denning).
(Image at top courtesy of NASA; image at bottom courtesy of Peter A.
Gnoffo and Jeffery A. White/NASA.).

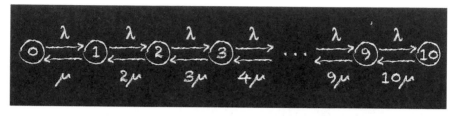

FIGURE 2 from "Computational Thinking in Science" (Denning)

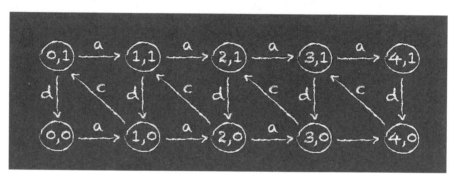

FIGURE 3 from "Computational Thinking in Science" (Denning)

Figure 1 from "Tangled Tangles" (Demaine, Demaine, and others).
Photo by Quanquan Liu, 2015.

Figure 2 from "Tangled Tangles" (Demaine, Demaine, and others)

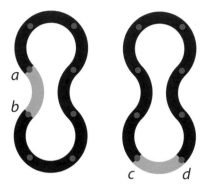

FIGURE 4 from "Tangled Tangles" (Demaine, Demaine, and others)

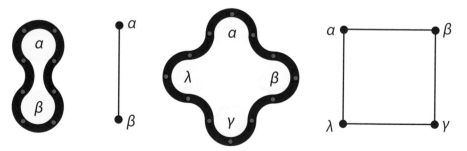

FIGURE 5 from "Tangled Tangles" (Demaine, Demaine, and others)

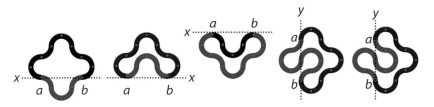

FIGURE 6 from "Tangled Tangles" (Demaine, Demaine, and others)

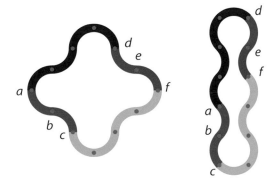

FIGURE 8 from "Tangled Tangles" (Demaine, Demaine, and others)

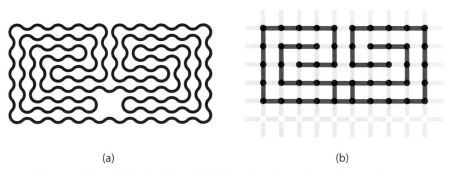

(a)   (b)

FIGURE 9 from "Tangled Tangles" (Demaine, Demaine, and others)

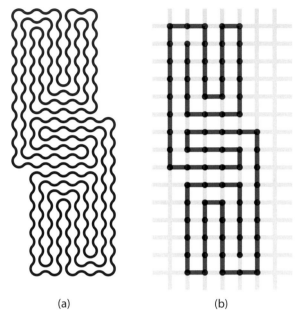

(a)   (b)

FIGURE 10 from "Tangled Tangles" (Demaine, Demaine, and others)

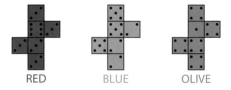

RED   BLUE   OLIVE

FIGURE 1 from "The Bizarre World of Nontransitive Dice" (Grime)

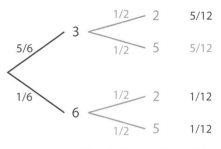

probability Red beats Blue = 7/12

FIGURE 2 from "The Bizarre World of Nontransitive Dice" (Grime)

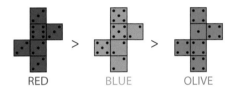

RED   >   BLUE   >   OLIVE

FIGURE 3 from "The Bizarre World of Nontransitive Dice" (Grime)

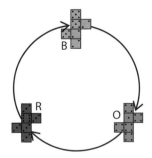

FIGURE 4 from "The Bizarre World of Nontransitive Dice" (Grime)

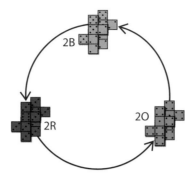

FIGURE 5 from "The Bizarre World of Nontransitive Dice" (Grime)

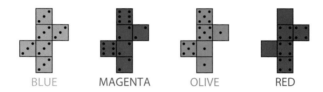

FIGURE 6 from "The Bizarre World of Nontransitive Dice" (Grime)

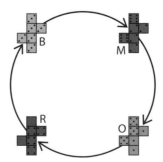

FIGURE 7 from "The Bizarre World of Nontransitive Dice" (Grime)

RED   BLUE   OLIVE   MAGENTA

FIGURE 10 from "The Bizarre World of Nontransitive Dice" (Grime)

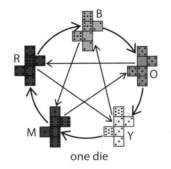

one die

FIGURE 11 from "The Bizarre World of Nontransitive Dice" (Grime)

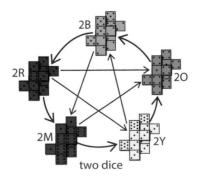

two dice

FIGURE 12 from "The Bizarre World of Nontransitive Dice" (Grime)

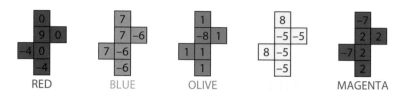

RED   BLUE   OLIVE   MAGENTA

FIGURE 14 from "The Bizarre World of Nontransitive Dice" (Grime)

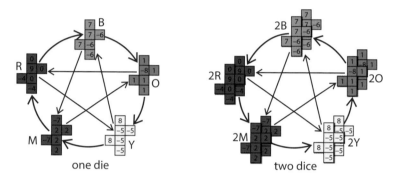

one die   two dice

FIGURE 15 from "The Bizarre World of Nontransitive Dice" (Grime)

FIGURE 2 from "The Bingo Paradox" (Benjamin/Kisenwether/Weiss)

FIGURE 3 from "The Bingo Paradox" (Benjamin/Kisenwether/Weiss)

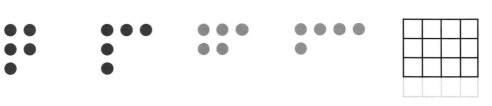

FIGURE 4 from "The Bingo Paradox" (Benjamin/Kisenwether/Weiss)

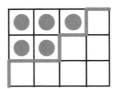

FIGURE 5 from "The Bingo Paradox" (Benjamin/Kisenwether/Weiss)

Figure 1 from "Written in Stone: The World's First Trigonometry Revealed in an Ancient Babylonian Tablet" (Mansfield/Wildberger). UNSW/Andrew Kelly, CC BY-SA.

Figure 4 from "Written in Stone: The World's First Trigonometry Revealed in an Ancient Babylonian Table" (Mansfield/Wildberger). UNSW/Andrew Kelly, Author provided.

The Writing Mathematician

Caroline Yoon

Writing isn't like math; in math, two plus two
always equals four no matter what your mood is like.
With writing, the way you feel changes everything.
—*Stephenie Meyer [1]*

I feel for my students when I hand them their first essay assignment. Many are mathematicians—students and teachers—who chose to study mathematics partly to avoid writing. But in my mathematics *education* courses, and in the discipline more generally, academic writing is part of our routine practice.

Mathematicians face some challenging stereotypes when it comes to writing. As the Meyer quote suggests, writing is often seen as ephemeral, subjective, and context-dependent, whereas mathematics is seen as enduring, universal, and context-free. Writing reflects self, mathematics transcends it; they are essentially unlike each other. This false dichotomy of writing versus mathematics can discourage mathematicians from writing, especially when combined with similarly coarse dichotomies such as *right brain* | *left brain, creativity* | *logic*, and *art* | *science*. Taken together, these dichotomies suggest that writing is outside the natural skill set of the mathematician, and that one's mathematics training not only neglects one's development as a writer, but also actively prevents it.

Where does this *writing* | *mathematics* dichotomy come from? It is profoundly unhelpful for our discipline of mathematics education: it assumes a gatekeeping role, turning away potential newcomers who might otherwise have a lot to offer. This essay deconstructs the *writing* | *mathematics* dichotomy by identifying similarities between the practices of academic writing and mathematics. I offer three writing-mathematics metaphors based on these similarities: writing as modeling, writing

TABLE 1. Metaphor Mappings between Mathematics and Writing Practice

Writing-Mathematics Metaphors	Unproductive Beliefs and Practices Addressed	Productive Beliefs and Practices Encouraged
Writing as modeling: The drafting process is like the modeling cycle, where early drafts are created in order to generate improved subsequent drafts.	Belief: One should know what to write before one starts writing. Practice: Wait until the last minute before beginning to write.	Belief: Writing can generate ideas. Practice: Create early drafts as a mechanism for figuring out what one wants to write.
Writing as problem solving: Writing is like solving a mathematical problem, where getting stuck is natural and expected.	Belief: If one knows what to write, the writing should flow easily. Practice: Give up when stuck.	Belief: Writing can be used to analyze and organize ideas. Practice: When stuck, approach one's writing metacognitively, and seek new ways of structuring one's attention to one's ideas.
Writing as proving: Academic writing is like a proof that performs a dialogic role in the way it addresses and seeks to convince a public.	Belief: Writing is a permanent, inert record of one's knowledge. Practice: Dismiss actual and potential reader interpretations of one's writing.	Belief: Writing is a dialogue with a public. Practice: Seek readers' interpretations of one's writing, from self-as-reader, imagined readers, and actual readers.

as problem solving, and writing as proving (Table 1). I propose that these metaphors might encourage students who identify more as mathematicians than writers to recognize and replace unproductive writing beliefs and practices with more productive ones that are grounded in familiar mathematical experiences.

I wish to delimit my intentions around this offering. I am not suggesting that mathematicians who read this essay will automatically become highly competent writers, simply by acknowledging similarities between writing and mathematics. Nor am I suggesting that one's skill in mathematical modeling, problem solving, and proving transfer effortlessly to the domain of writing. My goals are more modest and realistic. I want my students to work on their writing, just as they have worked on their mathematics. The three metaphors are offered as encouragement to *begin* this process, rather than being intended as a remedy or solution. I propose that mathematicians do not have to see themselves as starting from nothing when they engage in academic writing. Rather, they can view their writing development as building on competencies they have already honed in their mathematical training, but which they may not have formerly recognized as writerly.

The Imperfect Ideal: Writing as Modeling

Let us consider a prototypical mathematics education student who has spent weeks thinking, reading, and talking about her essay topic but only starts writing it the night before it is due. She writes one draft only—the one she hands in—and is disappointed with the low grade her essay receives. She wishes she had started earlier, but rationalizes to herself that she was still trying to figure out what she wanted to say up until the moment she started writing. It was only the pressure of the deadline that forced her to start writing; without it, she would have spent even more time thinking and reading to develop her ideas. After all, she reasons, there is no point writing when you do not know what to write about!

This "think first, write after" approach, sometimes known as the "writing up" model is a dangerous trap that many students fall into, and is at odds with the way writing often works (Menary 2007). The approach allows no room for an iterative drafting process, whereby imperfect drafts are written that are not intended as final copy, but are necessary steps toward subsequent improved drafts. Writing experts trade on the generative power of imperfect writing—they encourage writers to turn off their internal critic and allow themselves to write badly as a way of overcoming writing inertia and discovering new ideas (Elbow 1985). Lamott heralds the "shitty first draft" (1994) as an ideal (and achievable!) first goal in the writing process: anyone can produce a sketchy first draft that generates material that can be worked on, improved, and eventually rewritten into a more sharable form.

Mathematical modeling offers a compelling metaphor for the generative power of imperfect writing in the creative process. Like polished writing, polished mathematical models are seldom produced in a first attempt. A modeler typically begins with some understanding of the real [2] situation to be modeled. The modeler mathematizes variables and relationships from his or her understanding of the real situation, and "writes" it into an initial mathematical model. Next, the modeler runs the model to test it and interprets the results back into the real situation, comparing the mathematical output with real data. At this point, the modeler may notice information in the real situation that was previously missed, create a revised model based on this enhanced interpretation of the situation, and subject the model to further testing and revision, ultimately going through the full modeling cycle multiple times until she or he is satisfied (Borromeo-Ferri 2006). The model is his or her mathematical description of the situation, written in mathematical notation, and the modeler who publishes a mathematical model has typically created and discarded multiple models along the way, just as the writer who publishes a piece of writing has typically written and discarded multiple drafts along the way.

Novices to modeling might regard this iterative process as a waste of time: Why bother creating models that will only be discarded? The novice might try to bypass the iterations, thinking that if one thinks harder in the first place, one might produce a better model in an initial attempt. But the experienced modeler takes the more efficient approach of entering into modeling cycles, not avoiding them. Models self-propagate: the model one produces to express one's interpretation of a situation becomes a conceptual lens through which one can review the situation (Lesh and Doerr 2003); on this reviewing, one can notice deeper levels of structure that can be incorporated into a more powerful model. The modeler who views a real situation through multiple models typically notices more than one who spends the same amount of time trying to understand the real situation from an initial, untested point of view in the hope of producing a perfect first model. Similarly, the writer who writes, reads, and revises multiple drafts is likely to develop his or her ideas further than the writer who only thinks about his or her ideas in the hope of producing a perfect first draft.

A mathematical model is not valued for its verisimilitude, but for the opportunity it gives the user to manipulate, predict, or explain a system in a way that is not otherwise possible: "All models are wrong. But some are useful" (Box 1976). Modelers are aware that their mathematical descriptions exaggerate some features of a system and omit others, and they manipulate this exaggeration consciously to enhance the usefulness of the model. Perfection is simply not a relevant ideal in mathematical modeling; the utility of a model comes from its approximation of reality.

Modeling, like drafting, *expects* imperfection. This notion can be liberating: one can create a bad first model without worrying that it means one is a bad modeler. Modeling can be a useful metaphor for the generative role of writing for the mathematics education student who does not know what to write for her essay. Instead of waiting to figure out her ideas before writing, she can allow herself to write a bad first draft—full of contradictions, unfinished thoughts, ideas that are not structured well—in order to generate material that can be worked on and improved. Rather than viewing her bad first draft as evidence that she is a bad writer, she can view it as a useful tool for figuring out what she wants to say. She can enter into the drafting cycle knowing that the writing she does now will help her figure out what she wants to say.

The Thinking Laboratory: Writing as Problem Solving

The generation of ideas is just one of writing's roles; writing also plays a role in the analysis and organization of ideas that have been generated. Our prototypical student comes close to experiencing this second role when she knows what she wants to write about but struggles to write it. She has a thesis and outline for her intended essay, but the writing is painstaking. She spends hours writing the first sentence, only to delete it the next day. She throws up her hands and complains that she knows what she wants to say, but does not know how to say it—that she is "bad at writing" and that if she were "good at writing" she would be able to convey the same ideas more effectively, more eloquently. Good writers, she thinks, would not get stuck like this; their writing would flow elegantly from their pens.

It is an intimidating expectation. Even the most talented writer can be scared into inaction by the demand for elegant sentences created on the spot: "Be brilliant. Now!" Rather than finding it easy, many writers approach writing as an act of problem solving where getting stuck is a natural and expected part of the process. They may have a clear goal in mind but do not know how to get there. They write carefully, analyzing their writing as they go, shifting their attention back and forth between the writing that has been done and the goal that is beyond their reach. They scrutinize their inscriptions, using their writing as a thinking tool (Ong 1982) that helps them answer questions like these: "Should I start with this point, or leave it until these other two are developed first? Does this sentence move me toward my argument, or does it hint at a weakness that would undermine it? Is my central thesis workable, or should I modify it in light of the arguments in this first section?" Writers may begin with an outline plan, but they are not surprised when they abandon it as their analysis and writing lead them to reevaluate their goals and to create and work toward new ones.

Mathematical problem solving has been characterized as involving a similar process of working back and forth between givens and goals and posing new problems while exploring existing ones (Kontorovich et al. 2012). Mathematical problems require the solver to combine known ideas in previously unknown ways to create a solution method that was not already known. Problems have been distinguished from mathematical exercises (Schoenfeld 1985), which are strings of structurally similar questions that require students to practice a recently learned procedure. A student who has mastered a procedure will complete exercises quickly, easily, and accurately, much like the mythical "good writer" who issues a stream of perfect sentences in one sitting without breaking a sweat. This difference is important: Exercises require less mental effort because the script is already known; problems are more demanding, as they require the solvers to create the scripts as they go.

Writing an original essay is like trying to solve a problem—there is no script to follow; it must be created by simultaneously determining one's goals and figuring out how to achieve them. In both essay writing and mathematical problem solving, getting stuck is natural and expected—it is even a special kind of thrill. When mathematicians get stuck, they engage in metacognitive activity (Garofalo and Lester 1985), reviewing their desired goals and comparing them to their current ways of thinking in order to identify their domain of validity (Brousseau 2002)—the domain where their ways of thinking work and the point at which they break down. They also try to shift their ways of attending (Mason 2003) to the problem, often by using heuristics such as working backward, drawing a diagram, or solving a related problem (Polya 1945) in the hope of finding productive ways of thinking. They usually conduct such activities with a writing implement (pencil, chalk, whiteboard marker, keyboard) in hand, creating a "coded system of visible marks [. . .] whereby a writer could determine the exact words that a reader would generate from a given text" (Ong 1982, p. 83). That is, they write.

This last observation may come as a surprise to mathematicians who do not think of their problem-solving activity as writing. Doing mathematics—that is, the ordinary everyday concrete details of manipulating mathematical relationships and objects to notice new levels of structure and pattern—involves scratching out symbols, making marks, moving ideas around the page or board (Livingston 2015). To most mathematicians, writing is synonymous with "writing up one's results," and the semiotic activity that precedes the "writing up" is "working," "thinking," or "figuring things out," but not "writing" [3]. Of course, we are talking about different kinds of writing. The "writing up" of results is writing-as-reporting, performed at the end of the problem solving. In contrast, the semiotic activity that precedes "writing up" is a more analytic form of writing used to restructure one's ways of thinking during the problem-solving process. In academic writing, this writing-as-analysis is often performed without the expectation that it will appear intact in the final draft; it is often performed on scrap pieces of paper, with scribbles and arrows used to help the writer/thinker restructure his or her arguments.

Even if mathematicians do not call such semiotic activity writing, they readily acknowledge that the marks they make on the page, screen, or board are fundamental to their mathematical progress—that they help them analyze their ideas, that they "restructure thought" (Ong, 1982). My mathematician colleague Rod Gover once declared, "The whiteboard is our laboratory," when arguing for more space in university buildings. Another mathematician, Steven Galbraith, told my class, "I use a whiteboard when the ideas are too big for my head," conjuring an image of the whiteboard as an extension of the mind, with thinking distributed across multiple modalities. These marks help separate the known from the knower (Ong 1982), a distancing that allows the mathematician to "examine [one's] thoughts more consciously as a string of assertions in space" (Elbow 1985, p. 284) than if one spoke or thought them without writing them. Such distancing facilitates abstraction, which is a mathematical ideal.

Then why not call it writing? I suspect that the reluctance stems from a disciplinary tendency to overestimate the role of purely mental faculties and to underestimate the role of the diagrams, symbols, gestures, and glances that help mathematicians see new levels of structure. The archetypal story of mathematical discovery tells how Poincaré unexpectedly realized the link between non-Euclidean geometry and complex function theory while stepping on a bus, resumed his previous (nonmathematical) conversation for the rest of the bus ride with quiet certitude of his discovery, and only verified it when he got home—presumably with pen and paper (Hadamard 1945). The story creates a powerful image of pure thought (or divine inspiration) as the source of mathematical discovery, and it diminishes the role of writing to verification. Mathematicians' reluctance to call their analytical writing "writing" may also be motivated by a mathematical aesthetic that depends on the purity, simplicity, and abstraction of mathematics from the materiality of the so-called real world: ink and chalk dust are substantial, but the ideas they express endure and transcend. Perhaps this aesthetic prevents some mathematicians from acknowledging their role as writers who deal in messy physical marks, preferring instead to consider themselves thinkers reflecting on the ideal forms those marks express.

Why do I care that mathematicians acknowledge their semiotic activity as "writing"? Quite frankly, because they are good at it. They have spent years honing their ability to use writing to restructure their thoughts, to dissect their ideas, and identify new arguments—they possess an analytic discipline that most writers struggle with. Yet few of my mathematics education students take advantage of this in their academic writing. They want their writing to come out in consecutive, polished sentences, and become discouraged when it does not, rather than using their writing to analyze and probe their arguments as they do when they are stuck on mathematical problems. By viewing writing only as a medium for communicating perfectly formed thoughts, they deny themselves their own laboratories, their own thinking tools.

I am not suggesting that one's success in solving mathematical problems automatically translates into successful essay writing. But the metaphor of writing as problem solving may encourage a mathematics education student not to give up too easily when she finds herself stuck in her writing. Perhaps she can approach her writing metacognitively, reviewing her central thesis in light of her arguments, articulating her goals and identifying limitations in her current ways of thinking, just like she does when working on a mathematical problem. She may even try using problem-solving heuristics to break her current ways of attending to her writing and attend to her writing (and ideas) in new ways. At the very least, she may come to view her stuckness as a natural and expected state (Burton 2004), just as mathematicians do with mathematics, and recognize it as an opportunity to analyze and restructure her ideas.

Entering into Dialogue: Writing as Proving

Our prototypical student now has a good draft that she is happy with. She is satisfied that it represents her knowledge of the subject matter, and she has read extensively to check the accuracy of its content. A friend reads the draft and finds it difficult to understand. Unperturbed, our prototypical student attributes the reader's difficulty to insufficient subject knowledge. She is confident that her essay demonstrates her mastery of the topic, and that this will be recognized by her more knowledgeable teacher.

But the essay is not merely an inert record of a writer's knowledge, and its quality is not merely judged on the number of correct facts it contains. The essay is also a rhetorical act that seeks to engage a public. It has to do real work as a dialogic tool: it must address an audience; it must convince or persuade a public; it must inspire some kind of response or action. Writers take considerable care to shape their writing's addressivity (Bakhtin 1986) by seeking insight into their readers' experience of the writing.

Mathematical proofs are like expository essays in this regard: they must also convince an audience. When an undergraduate mathematics student switches from memorizing other peoples' proofs to constructing proofs of her own, she becomes concerned with how her proof (writing) performs under dialogic conditions. She is encouraged to test her proofs on different audiences: "convince yourself, convince a friend, convince an enemy" (Mason et al. 2010). She may take an undergraduate course in proof construction, in which she learns how to read and evaluate proofs as part of learning how to write and construct one (Weber 2010).

Mathematicians who construct proofs are like writers who create utterances as part of a dialogic chain in order to move, to persuade, "to do things to people" (Elbow 1985, p. 300). They are conscious of their audiences and often vary the style, amount of detail, and linguistic features of a proof (Pimm and Sinclair 2009), depending on whether it is being presented in real time to close colleagues, written in lecture notes for a graduate class, or submitted to a research journal. Moreover, these different modes of address can sometimes be generative: the awareness of different audiences' needs may alert the mathematician to new mathematical details or big picture isomorphisms that need to be worked out.

Mathematicians actively seek out listeners and readers who can identify weaknesses in the proofs they are constructing. Perhaps our prototypical student could enhance her writing by similarly seeking ways to evaluate her writing's dialogic performance. She could read her writing aloud to "hear it"; she could ask someone else to read it and tell her where their attention wanders; she could imagine herself writing for a particular person or audience; she could leave the essay and read it at a later date through the eyes of a reader. Such practices can result in productive tensions, where the reader sees something different from what the writer intended: "That's not what I meant!" And the writer reenters into dialogue, revising her argument, with a deeper understanding of what she is trying to say.

Reflections

I have challenged the common perception that writing is opposite to mathematics by offering three metaphors that highlight similarities between aspects of writing and mathematics practices (Table 1). According to Sword (2017), "metaphors, after all, are the stories that we tell ourselves about our relationship to the world. By changing our metaphors, we can rewrite our stories" (p. 191). I hope these metaphors will help students who identify more as mathematicians than writers to *re-story* themselves as writing mathematicians, who use writing as a tool for thinking more deeply about the mathematics and mathematics-education questions they care about most. I offer these metaphors with the conviction that it is ethically and fundamentally imperative that we support our mathematics education students' writing development—not only to help students reach their academic potential but also to ensure that our conversations include diverse voices (Geiger and Straesser, 2015) that are critical to our discipline's progress.

Acknowledgments

I wish to thank the writers, mathematicians, and mathematics educators who read and commented on drafts of this essay: John Adams, Benjamin Davies, John Griffith Moala, Igor' Kontorovich, Joel Laity, Jean-François Maheux, Roger Nicholson, Daniel Snell, James Sneyd, and Rebecca Turner, with special thanks to Sean Sturm, who first encouraged me to seek discipline-based metaphors for writing.

Notes

[1] http://stepheniemeyer.com/2008/08/midnight-sun.

[2] Readers may object to the use of the term "real"—is mathematics not real? Some writers on modeling us the term "extra-mathematical" instead of "real" to acknowledge the ontological slipstream; Borromeo-Ferri (2006) points out that some demarcation between mathematical and "real" or "extra-mathematical" worlds is necessary and the use of "real" to describe the "extra-mathematical" can be a pragmatic choice.

[3] See also Latour and Woolgar's (1979) anthropological observation of laboratory scientists as "compulsive and manic writers [. . .] who spend the greatest part of their day coding, marking, altering, correcting, reading, and writing" (pp. 48–49).

References

Bakhtin, M. (1986) *Speech Genres and other late Essays* (trans. McGee, V.). Austin, TX: University of Texas Press.

Borromeo-Ferri, R. (2006) Theoretical and empirical differentiations of phases in the modelling process. *ZDM* **38**(2), 86–95.

Box, G. E. (1976) Science and statistics. *Journal of the American Statistical Association*, **71**(356), 791–799.

Brousseau, G. (2002) *Theory of Didactical Situations in Mathematics: Didactique des Mathématiques*, 1970–1990 (Vol. 19). Springer Science & Business Media.

Burton, L. (2004) *Mathematicians as Enquirers: Learning about Learning Mathematics*. Norwell, MA: Kluwer Academic.

Elbow, P. (1985) The shifting relationships between speech and writing. *College Composition and Communication*, **36**(3), 283–303.

Garofalo, J., and Lester, F. K. (1985) Metacognition, cognitive monitoring, and mathematical performance. *Journal for Research in Mathematics Education*, **6**(3), 163–176.

Geiger, V., and Straesser, R. (2015) The challenge of publication for English non-dominant-language authors in mathematics education. *For the Learning of Mathematics*, **35**(3), 35–41.

Hadamard, J. (1945) *The Mathematician's Mind: The Psychology of Invention in the Mathematical Field*. Princeton, NJ: Princeton University Press.

Kontorovich, I., Koichu, B., Leikin, R., and Berman, A. (2012) An exploratory framework for handling the complexity of mathematical problem posing in small groups. *The Journal of Mathematical Behavior*, **31**(1), 149–161.

Lamott, A. (1994) *Bird by Bird: Some Instructions on Writing and Life*. New York: Pantheon Books.

Latour, B., and Woolgar, S. (1979) *Laboratory Life: The Social Construction of Scientific Facts*. Princeton, NJ: Princeton University Press.

Lesh, R., and Doerr, H. (2003) *Beyond Constructivism: Models and Modelling Perspectives on Mathematics Problem Solving, Learning, and Teaching*. Hillsdale, NJ: Lawrence Erlbaum Associates.

Livingston, E. (2015) The disciplinarity of mathematical practice. *Journal of Humanistic Mathematics*, **5**(1), 198–222.

Mason, J. (2003) On the structure of attention in the learning of mathematics. *The Australian Mathematics Teacher*, **59**(4), 17–25.

Mason, J., Burton, L., and Stacey, K. (2010) *Thinking Mathematically*. Harlow, U.K.: Pearson Education Ltd.

Menary, R. (2007) Writing as thinking. *Language sciences*, **29**(5), 621–632.

Ong, W. (1982) *Orality and Literacy: The Technologization of the Word*. London: Methuen.

Pimm, D., and Sinclair, N. (2009) Audience, style and criticism. *For the learning of mathematics*, **29**(2), 23–27.

Polya, G. (1945) *How to Solve it: A new aspect of Mathematical Method*. Princeton, NJ: Princeton University Press.

Schoenfeld, A. H. (1985) *Mathematical Problem Solving*. Orlando, FL: Academic Press, Inc.

Sword, H. (2017). *Air & Light & Time & Space: How Successful Academics Write*. Cambridge MA: Harvard University Press.

Weber, K. (2010) Mathematics majors' perceptions of conviction, validity, and proof. *Mathematical Thinking and Learning*, **12**(4), 306–336.

Contributors

Chris Arney is a professor of mathematics at the U.S. Military Academy at West Point, NY, where he teaches undergraduate mathematics and network science and models cooperative systems, information networks, pursuit-evasion systems, intelligence processing, artificial intelligence, and language for robots. His work in number theory was published in a paper coauthored with Paul Erdős and Joe Arkin entitled "Two Theorems of Arkin-Arney-Erdős" in *Congressus Numerantium* (1996). Arney is the founding director of the annual international undergraduate contest Interdisciplinary Contest in Modeling, which celebrated its twentieth year in 2018.

Michael J. Barany studies the institutional, political, social, conceptual, and material dimensions of abstract knowledge in modern societies. From 2016-2018 he was a postdoctoral fellow in the Dartmouth College Society of Fellows, and from 2018 is Lecturer in the History of Science at the University of Edinburgh. His Ph.D. dissertation, "Distributions in Postwar Mathematics," received a special mention from the history division of the International Union for the History and Philosophy of Science and Technology among dissertations completed between 2012 and 2016. This is his third appearance in the *Best Writing on Mathematics* series. You can find these and many more essays at http://mbarany.com and follow him on twitter @mbarany.

Arthur Benjamin is the Smallwood Family Professor of Mathematics at Harvey Mudd College. His research interests include combinatorics and number theory, with a special fondness for Fibonacci numbers. Many of these ideas appear in his book (coauthored with Jennifer Quinn), *Proofs That Really Count: The Art of Combinatorial Proof*, which received the Beckenbach Book Prize from the Mathematical Association of America (MAA). Professors Benjamin and Quinn were the editors of *Math Horizons* magazine from 2004 through 2008. In 2000, he received the Haimo Award for Distinguished Teaching from the MAA, and he has served as the MAA's Polya Lecturer from 2006 to 2008. Dr. Benjamin is also a professional magician who performs his mixture of math and magic to audiences all over the world. He has given three TED talks, which have been viewed more than 15 million times. In 2017, he received the Communications Award from the Joint Policy Board of Mathematics.

Benjamin Braun is an associate professor in the Department of Mathematics at the University of Kentucky. His mathematical research is in geometric and algebraic combinatorics, and he is active in mentoring graduate and undergraduate students. His scholarly interests in teaching and learning include active learning, using writing in mathematics courses, preservice teacher education, pedagogical use of the history of mathematics, and connections between mathematics education and other disciplines.

Priscilla Bremser is the Nathan Beman Professor of Mathematics at Middlebury College. She earned her Ph.D. at Johns Hopkins University. Her research interests are number theory, especially the theory of finite fields, and mathematics education at all levels. She is an instructor at the Vermont Mathematics Initiative, a master's degree program for practicing teachers, an experience that has informed her own teaching as well as her understanding of precollege teaching and learning.

Erik D. Demaine is a computer scientist and an artist. Since 2001, he has been a professor in computer science at the Massachusetts Institute of Technology. Demaine's research interests range throughout algorithms, from data structures for improving web searches to the geometry of understanding how proteins fold the computational difficulty of playing games. He received a MacArthur Fellowship (2003) as a "computational geometer tackling and solving difficult problems related to folding and bending—moving readily between the theoretical and the playful, with a keen eye to revealing the former in the latter." Demaine cowrote a book about the theory of folding, together with Joseph O'Rourke (*Geometric Folding Algorithms*, 2007) and a book about the computational complexity of games, together with Robert Hearn (*Games, Puzzles, and Computation*, 2009). Together with his father Martin, his interests span the connections between mathematics and art, including curved origami sculptures in the permanent collections of the Museum of Modern Art in New York and the Renwick Gallery, part of the Smithsonian Institution.

Martin L. Demaine is an artist and a computer scientist. He started the first private hot glass studio in Canada and has been called the father of Canadian glass. Since 2005, he has been the Angelika and Barton Weller Artist-in-Residence at the Massachusetts Institute of Technology. Demaine works together with his son Erik in paper, glass, and other material. They use their exploration in sculpture to help visualize and understand unsolved problems in mathematics, and their scientific abilities to inspire new art forms. Their artistic work includes curved origami sculptures in the permanent collections of the Museum of Modern Art in New York, and the Renwick Gallery, part of the Smithsonian Institution. Their scientific work includes more than sixty published joint papers, including several about combining mathematics and art. They recently won a Guggenheim Fellowship (2013) for exploring folding of other materials, such as hot glass.

Peter J. Denning began building electronic circuits as a teenager. His computer built from pinball machine parts won the science fair in 1959, launching him into the new field of computing. At MIT for his doctorate in 1968, he worked on Multics, a precursor of today's "cloud computing" systems. He has taught computer science at Princeton, Purdue, and George Mason universities, and the Naval Postgraduate School in Monterey, CA. A pioneer in operating systems and computer networks, he invented the "working set," a way of managing memory for optimal system throughput. From directing a lab at NASA's Ames Research Center, he wrote *The Innovator's Way* (MIT Press, 2010) on leadership practices to generate adoption of innovations. He also published *Great Principles of Computing* (MIT Press, 2015). He has won twenty-eight awards for his work in computing science and education. He is a past president of the Association for Computing Machinery, the oldest scientific society in computing. He is currently editor of *Ubiquity* (ubiquity.acm.org).

Robbert Dijkgraaf is the director and Leon Levy Professor of the Institute for Advanced Study, one of the world's leading centers for curiosity-driven basic research in the sciences and humanities. Past President of the Royal Netherlands Academy of Arts and Sciences, Dijkgraaf is a mathematical physicist who has made important contributions to string theory and the advancement of science education. He is a recipient of the Spinoza Prize, the highest scientific award in the Netherlands, and is a Knight of the Order of the Netherlands Lion. Dijkgraaf is most recently coauthor with Abraham Flexner of *The Usefulness of Useless Knowledge* (Princeton University Press, 2017).

Art M. Duval is a professor in the Department of Mathematical Sciences at the University of Texas at El Paso. He earned his Ph.D. in mathematics from the Massachusetts Institute of Technology, and his primary research interests are in algebraic and topological combinatorics. He has been involved with the preparation of K–12 mathematics teachers in one way or another for most of his career and has worked with local school districts and the El Paso Collaborative for Academic Excellence on issues of vertical and horizontal alignment.

José Ferreirós is a professor of logic and the philosophy of science at the University of Seville, Spain. A former Fulbright Fellow at UC Berkeley, and member of the *Académie Internationale de Philosophie des Sciences*, he was founding member and first president of the Association for the Philosophy of Mathematical Practice. Among his publications, one finds a history of set theory and its role in modern math, *Labyrinth of Thought* (Birkhäuser, 1999), *Mathematical Knowledge and the Interplay of Practices* (Princeton University Press, 2016), an intellectual biography of Riemann (*Riemanniana Selecta*, CSIC, Madrid, 2000), and *The Architecture of Modern Mathematics* (Oxford University Press, 2006). He has three daughters and loves music.

James Grime currently runs the Enigma Project and travels the United Kingdom and the world giving talks about the history and mathematics of codes and code breaking. Grime earned a Ph.D. in mathematics from the University of York working on combinatorial representation theory. He then joined the Millennium Mathematics Project, an education and outreach project from the University of Cambridge. Grime also creates mathematical toys for Maths Gear and often appears on the YouTube channel Numberphile, which has viewers all over the world and now has more than 2 million subscribers.

Adam Hesterberg is a Ph.D. student in the MIT Mathematics Department, studying computational geometry and computational complexity. He graduated summa cum laude from Princeton with an A.B. in mathematics after writing a thesis on graph minors. He often teaches at Canada/USA Mathcamp, a mathematical summer program for high schoolers.

Marijn J. H. Heule is a research assistant professor at the University of Texas at Austin who received his Ph.D. at Delft University of Technology (2008). His contributions to satisfiability (SAT) solving have enabled him and others to solve hard problems in formal verification and mathematics. He has developed award-winning SAT solvers, and his preprocessing and proof-producing techniques are used in many state-of-the-art solvers. Heule is one of the editors of the *Handbook of Satisfiability* and has coorganized the SAT competitions in recent years. Four of his recent publications were awarded best paper at the SAT (International Conference on Theory and Applications of Satisfiability Testing), HVC (Haifa Verification Conference), CADE (Conference on Automated Deduction), and TACAS (Tools and Algorithms for the Construction and Analysis of Systems) conferences.

Joseph Kisenwether earned undergraduate degrees in mathematics from New Jersey Institute of Technology in 1994 and education from Kean

University in 1996. He spent several years as a high school math teacher before finding his calling in the casino industry. Since founding Craftsman Gaming in 2012, he has been working as an independent mathematics and game design consultant.

Nancy Emerson Kress is a doctoral student in mathematics curriculum and instruction in the School of Education at the University of Colorado at Boulder. She is a recipient of the Chancellor's Award for Excellence in STEM Education at the University of Colorado at Boulder. She taught high school mathematics for 20 years and holds an M.Ed. and Master's of Science for Teachers (M.S.T.) degree in mathematics. Kress's primary research interests focus on structures that support participation in mathematics by students who are members of underrepresented groups in math.

Oliver Kullmann is an associate professor at Swansea University, United Kingdom. His main research topic is the development of the emerging theory of satisfiability (SAT). His Ph.D. thesis (University of Frankfurt) was based on the development of the method for worst-case analysis of SAT algorithms, often referred to as "measure-and-conquer." His *Handbook of Satisfiability* chapter "Fundaments of Branching Heuristics" combines this theoretical approach with his practical experience as a SAT solver and author. After that, he embarked on theoretical investigations into structural aspects of SAT, notably autarky theory (*Handbook of Satisfiability* Chapter "Minimal Unsatisfiability and Autarkies"), and various practically motivated aspects of complexity theory. Recent publications obtained best-paper awards at HVC (Haifa Verification Conference), SOFSEM (International Conference on Current Trends in Theory and Practice of Computer Science), and SAT (International Conference on Theory and Applications of Satisfiability Testing).

Quanquan Liu is a graduate student in theoretical computer science at MIT. She received her bachelor's degree in computer science and mathematics in 2015, also from MIT. During her undergrad years, she performed research in theoretical computer science, computer systems, and theoretical biology and chemistry. Now, her interests lie in graph algorithms, approximation algorithms, cache-efficient algorithms, hardness of approximation, data structures, and memory-hard functions. In the past, she has been a coach for the USA Computing Olympiad summer camp and a tutor for the Women's Technology Program at MIT.

Elise Lockwood is an assistant professor in the Mathematics Department at Oregon State University. She received her Ph.D. in mathematics education from Portland State University and was a postdoctoral scholar at the University of Wisconsin—Madison. Her primary research interests focus on undergraduate students' reasoning about combinatorics, and she is passionate about improving the teaching and learning of discrete mathematics. In 2017, she was awarded an NSF Career Award, in which she seeks to investigate ways that computational activities can be leveraged to help students solve counting problems more successfully.

Daniel Mansfield is a lecturer in the School of Mathematics and Statistics, UNSW Sydney, with research interests in ergodic theory and the history of mathematics. Mansfield is a passionate supporter of mathematics education, and his excellence in teaching has been recognized by several awards, including the 2017 KPMG Inspiring Teacher Award in a First Year Undergraduate Program.

Lucy H. Odom is an adjunct instructor of mathematics at Santa Monica College and at Long Beach City College. She earned her master's degree in mathematics at San Francisco State University in 2015. Her master's thesis is titled "An Overlap Criterion for the Tiling Problem of the Littlewood Conjecture." The thesis concerned the study of the geometry of certain three-dimensional lattices and the corresponding polygon-shaped tiles in the plane that arise in the study of rational approximations to a pair of real numbers. She received the College of Science and Engineering ARCS Scholarship for two years in a row, 2013–2015, which helped to fund her graduate research work. Her current professional interests include mathematics education, especially at the precollegiate level.

Isabel M. Serrano is pursuing an undergraduate degree in applied computational mathematics and a minor in history at California State University, Fullerton. During her undergraduate career, she published the article, "A Medieval Mystery: Nicole Oresme's Concept of *Curvitas*" alongside Bogdan Suceavă in the American Mathematical Society's *Notices* and the *2016 Best Writing on Mathematics*. In addition, Serrano has participated in NIH's Maximizing Access to Research Careers program, where she worked to build a mathematical model for the 2016 Zika virus outbreak in Puerto Rico. Moving forward, she plans to pursue a Ph.D. in computational biology at the University of California, Berkeley.

Francis Edward Su is the Benediktsson-Karwa Professor of Mathematics at Harvey Mudd College, and past president of the Mathematical Association of America. He is equally fascinated by research questions in topological combinatorics and by the reflection necessary for communicating mathematics effectively to any audience. He received the 2001 MAA Hasse Prize for expository writing and the 2013 MAA Haimo Award for distinguished teaching. Su is the creator of the popular *Math Fun Facts* website and *MathFeed*, the math news app. His book building on his article "Mathematics for Human Flourishing" will be published by Yale University Press in 2019. He's on Twitter at @mathyawp.

Bogdan D. Suceavă is a professor of mathematics at California State University, Fullerton. His academic interests include differential geometry, metric geometry, and the history of mathematics. His research work has appeared in the *American Mathematical Monthly, Differential Geometry and Its Applications, Houston Journal of Mathematics,* the *Mathematical Intelligencer, Results in Mathematics, Taiwanese Journal of Mathematics, Elemente der Mathematik,* and other journals. Suceavă is also the author of several novels, some available in English: *Coming from an Off-Key Time* (Northwestern University Press, 2011); *Miruna, a Tale* (Twisted Spoon Press, 2014). His most recent novel in Romanian is *The Republic* (Polirom Press, 2016), and his most recent volume is the essay *A History of Lapses* (in Romanian) (Polirom Press, 2017).

Ron Taylor is a professor of mathematics at Berry College in Georgia and a Project NExT Fellow of the Mathematical Association of America. He earned his Ph.D. in mathematics from Bowling Green State University in 2000. As a researcher, he considers himself to be a dabbler, having done work in functional analysis and operator theory, knot theory, geometry, number theory, symbolic logic, graph theory, and recreational mathematics. Taylor is coauthor, with Patrick X. Rault, of the MAA textbook *A TEXas Style Introduction to Proof.* He has received several teaching awards, including the 2018 Haimo Award for Distinguished Teaching, presented by the MAA.

Robert S. D. Thomas (http://orcid.org/0000-0003-4697-4209) is a fellow of St. John's College and emeritus professor in the mathematics department at the University of Manitoba, Winnipeg, Canada. Being a member of Common Room at Wolfson College, Oxford, is helpful. His main activity is editing *Philosophia Mathematica*, the philosophical journal of the Canadian Society for History and Philosophy of Mathematics, of which he is a former president, but he also does research on geometry, not necessarily pure, and its history. For example, *Euclid's* Phaenomena: *A Translation and Study of a Hellenistic Treatise in Spherical Astronomy,* with J. L. Berggren. This is his second appearance in this book series. The first was "Mathematics is not a game, but . . ." from *The Mathematical Intelligencer,* where his first appearance was "Mathematics and Narrative." He has been jogging in moderation since 1969.

Ryuhei Uehara is a full professor in the school of information science, Japan Advanced Institute of Science and Technology (JAIST). He received B.E., M.E., and Ph.D. degrees from the University of Electro-Communications, Japan, in 1989, 1991, and 1998, respectively. He was a researcher at Canon Inc. from 1991 to 1993. In 1993, he joined Tokyo Woman's Christian University as an assistant professor. He was a lecturer during 1998–2001 and an associate professor during 2001–2004 at Komazawa University. He moved to JAIST in 2004. His research interests include computational complexity, algorithms and data structures, and graph algorithms. Uehara is especially engrossed in computational origami, games, and puzzles from the viewpoints of theoretical computer science. He is a member of the Information Processing Society of Japan, the *Institute of Electronics, Information and Communication Engineers*, and the European Association for Theoretical Computer Science.

Ben Weiss is a Harvey Mudd College alum and is currently a senior software engineer at Google. He is the developer of the app Frax, which allows users to navigate fractal images in real time. He is a three-time member of Team USA in the extreme sport of freediving and has held his breath for more than seven minutes. He lives in Southern California, where he and his partner Jenna recently welcomed their new daughter, Aurora.

Margaret Wertheim is a writer whose work focuses on relations between science and the wider cultural landscape. Her books include *The Pearly Gates of Cyberspace: A History of Space from Dante to the Internet* and *Pythagoras' Trousers: God, Physics and the Gender War*. As a science communicator, she has received the 2016 Klopsteg Memorial Award from the American Association of Physics Teachers and the 2017 Scientia Medal in Australia. Her essays have been included in *Best American Science Writing* and *Best Australian Science Writing*. Wertheim also creates projects at the intersection of science and art, including the Crochet Coral Reef, a worldwide endeavor that marries handicraft and hyperbolic geometry to create giant sculptural installations of living reefs.

Diana White is an associate professor of mathematics and mathematics education at the University of Colorado Denver and the director of the National Association of Math Circles. She earned her Ph.D. from the University of Nebraska and was a postdoctoral scholar at the University of South Carolina. Her current research and scholarly interests include teacher training and teacher professional development, using primary sources to teach undergraduate mathematics, and mathematical outreach for K–12 students.

Norman Wildberger of the School of Mathematics and Statistics, UNSW Sydney, has worked in harmonic analysis, representation theory of Lie groups, hypergroups and universal hyperbolic geometry, and is an award-winning lecturer. Wildberger wrote the foundational book *Divine Proportions: Rational Trigonometry to Universal Geometry*, which introduced rational trigonometry. He also has a popular YouTube math channel called Insights into Mathematics, where he explores the foundations of mathematics and new directions for math education.

Peter Winkler is the William Morrill Professor of Mathematics and Computer Science at Dartmouth College. He is the author of about 150 research papers and holds a dozen patents in marine navigation, cryptography, holography, gaming, optical networking, and distributed computing. His research is primarily in combinatorics, probability, and the theory of computing, with forays into statistical physics. He is a winner of the Mathematical Association of America's Lester R. Ford Award and the David P. Robbins Prize. Winkler has also written two collections of mathematical puzzles, a book on cryptology in the game of bridge, and a portfolio of compositions for ragtime piano. He's working on a new puzzle book.

Caroline Yoon is an associate professor at the University of Auckland, New Zealand. She received her undergraduate and master's degrees in mathematics from the department where she now works and earned a Ph.D. in mathematics education and learning sciences from Indiana University, Bloomington. She researches and teaches mathematics education and studies mathematical insight, task design, and writing.

Notable Writings

Abtahi, Yasmine, Mellony Graven, and Stephen Lerman. "Conceptualising the More Knowledgeable Other within a Multi-Directional ZPD." *Educational Studies in Mathematics* 96(2017): 275–87.

Ancker, Jessica S., and Melissa D. Begg. "Using Visual Analogies to Teach Introductory Statistical Concepts." *Numeracy* 10.2(2017).

Andrews, Noam. "Tabula III: Kepler's Mysterious Polyhedral Model." *Journal for the History of Astronomy* 48.3(2017): 281–311.

Andrews, R. J., and Howard Wainer. "The Great Migration: A Graphics Novel." *Significance* 14.5(2017): 14–19.

Aristidou, Michael. "Some Thoughts on the Epicurean Critique of Mathematics." *Journal of Humanistic Mathematics* 7.2(2017).

Aubin, Sean, Aaron R. Voelker, and Chris Eliasmith. "Improving with Practice: A Neural Model of Mathematical Development." *Topics in Cognitive Science* 10(2017): 6–20.

Ayache, Elie. "Time and Black–Scholes–Merton." *Wilmott Magazine* 49(2017): 24–33.

Bair, Jacques. "Interpreting the Infinitesimal Mathematics of Leibniz and Euler." *Journal for General Philosophy of Science* 48(2017): 195–238.

Baker, Alan. "Mathematical Spandrels." *Australian Journal of Philosophy* 95.4(2017): 779–93.

Bandle, Catherine. "Dido's Problem and Its Impact on Modern Mathematics." *Notices of the American Mathematical Society* 64.9(2017): 980–84.

Barahmand, Ali. "The Boundary between Finite and Infinite States through the Concept of Limits of Sequences." *International Journal of Science and Mathematics Education* 15(2017): 569–85.

Barany, Michael J., Anne-Sandrine Paumier, and Jesper Lützen. "From Nancy to Copenhagen to the World: The Internationalization of Laurent Schwartz and His Theory of Distributions." *Historia Mathematica* 44.4(2017): 367–94.

Bardawil, Fadi, et al. "Immigration, Freedom, and the History of the Institute for Advanced Study." *Notices of the American Mathematical Society* 64.10(2017): 1060–68.

Barford, Megan. "D.176: Sextants, Numbers, and the Hydrographic Office of the Admiralty." *History of Science* 55.4(2017): 431–56.

Barka, Zoheir. "The Hidden Symmetries of the Multiplication Table." *Journal of Humanistic Mathematics* 7.1(2017): 189–203.

Bass, Hyman. "Designing Opportunities to Learn Mathematics Theory-Building Practices." *Educational Studies in Mathematics* 95(2017): 229–44.

Beeley, Philip. "'To the publike advancement': John Collins and the Promotion of Mathematical Knowledge in Restoration England." *BSHM Bulletin: Journal of the British Society for the History of Mathematics* 32.1(2017): 61–74.

Belot, Gordon. "Curve-Fitting for Bayesians?" *British Journal for the Philosophy of Science* 68(2017): 689–702.

Bender, Andrea, and Sieghard Beller. "The Power of 2: How an Apparently Irregular Numeration System Facilitates Mental Arithmetic." *Cognitive Science* 41(2017): 158–87.

Bernstein, Mira, and Moon Duchin. "A Formula Goes to Court: Partisan Gerrymandering and the Efficiency Gap." *Notices of the American Mathematical Society* 64.9(2017): 1020–24.

Biggs, Norman. "More Seventeenth-Century Networks." *BSHM Bulletin: Journal of the British Society for the History of Mathematics* 32.1(2017): 30–39.

Bland, Andre, et al. "Happiness Is Integral but Not Rational." *Math Horizons* 25.1(2017): 8–11.

Blåsjö, Viktor. "On What Has Been Called Leibniz's Rigorous Foundation of Infinitesimal Geometry by Means of Riemannian Sums." *Historia Mathematica* 44.2(2017): 134–49 [with commentaries in successive journal issues].

Błaszczyk, Piotr, et al. "Controversies in the Foundations of Analysis: Comments on Schubring's 'Conflicts.'" *Foundations of Science* 22(2017): 125–40.

Błaszczyk, Piotr, et al. "Is Leibnizian Calculus Embeddable in First Order Logic?" *Foundations of Science* 22(2017): 717–31.

Błaszczyk, Piotr, et al. "Toward a History of Mathematics Focused on Procedures." *Foundations of Science* 22(2017): 763–83.

Boehm, Hans-J. "Small-Data Computing: Correct Calculator Arithmetic." *Communications of the ACM* 60.8(2017): 44–49.

Boldyrev, Ivan, and Olessia Kirtchik. "The Cultures of Mathematical Economics in the Postwar Soviet Union: More Than a Method, Less Than a Discipline." *Studies in History and Philosophy of Science, A* 63(2017): 1–10.

Booß-Bavnbek, Bernhelm. "Mathematics: Easy *and* Hard. Why?" *The Mathematical Intelligencer* 39.3(2017): 61–72.

Bossaerts, Peter, and Carsten Murawski. "Computational Complexity and Human Decision-Making." *Trends in Cognitive Sciences* 21.12(2017): 917–29.

Brenton, Lawrence. "A Bigger Altar: Geometry and Ritual." *Math Horizons* 25.1(2017): 8–11.

Brown, Nicholas J. L., and James C. Coyne. "Emodiversity: Robust Predictor of Outcomes or Statistical Artifact?" *Journal of Experimental Psychology: General* 146.9(2017): 1372–77.

Bruderer, Herbert. "Computing History beyond the U.K. and U.S.: Selected Landmarks from Continental Europe." *Communications of the ACM* 60.2(2017): 76–84.

Buijsman, Stefan. "Accessibility of Reformulated Mathematical Content." *Synthese* 194(2017): 2233–50.

Bunge, Mario. "Why Axiomatize?" *Foundations of Science* 22(2017): 695–707.

Calude, Cristian S., and Giuseppe Longo. "The Deluge of Spurious Correlations in Big Data." *Foundations of Science* 22(2017): 595–612.

Campbell, J. M. "Visualizing Large-Order Groups with Computer-Generated Cayley Tables." *Journal of Mathematics and the Arts* 11.2(2017): 67–99.

Campbell, Paul J. "Coding for All." *UMAP Journal* 37.4(2017): 333–37.

Cao, Longbing. "Data Science: Challenges and Directions." *Communications of the ACM* 60.8(2017): 59–68.

Capaldi, Mindy, and Tiffany Kolba. "Carcassonne in the Classroom." *The College Mathematics Journal* 48.4(2017): 265–73.

Capobianco, Giovanni, Maria Rosaria Enea, and Giovanni Ferraro. "Geometry and Analysis in Euler's Integral Calculus." *Archive for History of Exact Sciences* 71(2017): 1–38.

Castelvecchi, Davide. "Long-Awaited Mathematics Proof Could Help Scan Earth's Innards." *Nature Online* Feb 10, 2017.

Cellucci, Carlo. "Is Mathematics Problem Solving or Theorem Proving?" *Foundations of Science* 22(2017): 183–99.

Cetin, Ibrahim, and Ed Dubinsky. "Reflective Abstraction in Computational Thinking." *The Journal of Mathematical Behavior* 47(2017): 70–80.

Chemla, Karine. "Abstraction as a Value in the Historiography of Mathematics in Ancient Greece and China." In *Ancient Greece and China Compared*, edited by G. E. R. Lloyd and Jingyi Jenny Zhao, Cambridge, U.K.: Cambridge University Press, 2018, 290–325.

Chen, Zhihui. "Scholars' Recreation of Two Traditions of Mathematical Commentaries in Late Eighteenth-Century China." *Historia Mathematica* 44.2(2017): 105–33.

Chernoff, Egan J. "Numberlines: Hockey Line Nicknames Based on Jersey Numbers." *Mathematics Enthusiast* 14.1–3(2017): 371–85.

Coleman, John A. "Groups and Physics: Dogmatic Opinions of a Senior Citizen." *Notices of the AMS* 64.1(2017): 8–17.

Cone, Randall E. "Perchance to Dream: Art, Mathematics, and Shakespeare." *Journal of Humanistic Mathematics* 7.2(2017).

Cook, Susan Wagner, et al. "Hand Gesture and Mathematics Learning: Lessons from an Avatar." *Cognitive Science* 41(2017): 518–35.

Cormode, Graham. "Data Sketching." *Communications of the ACM* 60.9(2017): 48–55.

Corry, Leo. "Turing's Pre-War Analog Computers: The Fatherhood of the Modern Computer Revisited." *Communications of the ACM* 60.8(2017): 50–58.

Cragg, Lucy, et al. "Direct and Indirect Influences on Executive Functions on Mathematical Achievement." *Cognition* 162(2017): 12–26.

Crannell, Annalisa, Marc Frantz, and Fumiko Futamura. "The Image of a Square." *The American Mathematical Monthly* 124.2(2017): 99–115.

Craver, Carl L., and Mark Povich. "The Directionality of Distinctively Mathematical Explanations." *Studies in History and Philosophy of Science, A* 63(2017): 31–38.

Crespo, Ricardo, and Fernando Tohmé. "The Future of Mathematics in Economics." *Foundations of Science* 22(2017): 677–93.

Crilly, Tony, Steven H. Weintraub, and Paul R. Wolfson. "Arthur Cayley, Robert Harley and the Quintic Equation: Newly Discovered Letters 1859–1863." *Historia Mathematica* 44.2(2017): 150–69.

Dawkins, Paul Christian, and Keith Weber. "Values and Norms of Proof for Mathematicians and Students." *Educational Studies in Mathematics* 95(2017): 123–42.

Degrande, Tine, Lieven Verschaffe, and Wim Van Dooren. "Spontaneous Focusing on Quantitative Relations: Towards a Characterization." *Mathematical Thinking and Learning* 19.4(2017): 260–75.

DeLiema, David. "Co-Constructed Failure Narratives in Mathematics Tutoring." *Instructional Science* 45(2017): 709–35.

Denning, Peter J. "Remaining Trouble Spots with Computational Thinking." *Communications of the ACM* 60.6(2017): 33–39.

Denning, Peter J., and Ted G. Lewis. "Exponential Laws of Computing Growth." *Communications of the ACM* 60.1(2017): 54–65.

Denning, Peter J., Matti Tedre, and Pat Yongpradit. "Misconceptions about Computer Science." *Communications of the ACM* 60.3(2017): 31–33.

Dewar, Jacqueline M. "Women and Mathematics: A Course and a Scholarly Investigation." *BSHM Bulletin: Journal of the British Society for the History of Mathematics* 32.3(2017): 246–53.

Donaldson, Thomas. "The (Metaphysical) Foundations of Arithmetic?" *Noûs* 51.4(2017): 775–801.

Dover, Jeremy M. "Cybersecurity and the Mathematically-Inclined." *Pi Mu Epsilon Journal* 14.6(2017): 357–64.

Ducheyne, Steffen. "Different Shades of Newton: Herman Boerhaave on Newton *Mathematicus*, *Philosophus*, and *Optico-Chemicus*." *Annals of Science* 74.2(2017): 108–25.

Edgar, Tom, and N. Chris Meyer. "A Visual Validation of Viète's Verification." *The College Mathematics Journal* 48.2(2017): 90–96.

Elgersma, Michael, and Stan Wagon. "An Asymptotically Closed Loop of Tetrahedra." *The Mathematical Intelligencer* 39.3(2017): 40–45.

Ely, Robert. "Definite Integral Registers Using Infinitesimals." *The Journal of Mathematical Behavior* 48(2017): 152–67.

Engelson, Morris. "The Biblical Value of Pi in Light of Traditional Judaism." *Journal of Humanistic Mathematics* 7.2(2017).

Ernst, Dana C., Angie Hodge, and Stan Yoshinobu. "What Is Inquiry-Based Learning?" *Notices of the AMS* 64.6(2017): 570–74.

Etz, Alexander, and Eric-Jan Wagenmakers. "J.B.S. Haldane's Contribution to the Bayes Factor Hypothesis Test." *Statistical Science* 32.2(2017): 313–29.

Farley, Rosemary, et al. "Modeling-First Approach to Teaching Differential Equations." *UMAP Journal* 37.4(2017): 381–406.

Feldman, Jacob. "What Are the 'True' Statistics of the Environment?" *Cognitive Science* 41(2017): 1871–1903.

Fiocca, Alessandra. "The *Bullettino di Bibliografia e di Storia delle Scienze Matematiche e Fisiche* (1868–1887), an Example of the Internationalization of Research." *Historia Mathematica* 44.1(2017): 1–30.

Fitzherbert, John. "Ghosts of Mathematicians Past: Bharati Krishna and Gabriel Cramer." *Australian Senior Mathematics Journal* 31.1(2017): 48–58.

FitzSimons, Gail E., and Lisa Björklund Boistrup. "In the Workplace Mathematics Does Not Announce Itself: Towards Overcoming the Hiatus between Mathematics Education and Work." *Educational Studies in Mathematics* 95(2017): 329–49.

Franklin, James. "Discrete and Continuous: A Fundamental Dichotomy in Mathematics." *Journal of Humanistic Mathematics* 7.2(2017).

Frazier, Shelby. "A Brief Look at the Evolution of Modeling Hyperbolic Space." *Mathematics Enthusiast* 14.1–3(2017): 29–51.

Freeman, David M. "Epicycloid Curves and Continued Fractions." *Journal of Mathematics and the Arts* 11.2(2017): 100–113.

Fukawa-Connelly, Timothy, Keith Weber, and Juan Pablo Mejía-Ramos. "Informal Content and Student Note-Taking in Advanced Mathematics Classes." *Journal for Research in Mathematics Education* 48.5(2017): 567–79.

Gage, Joseph. "Undergraduate Algebra in Nineteenth-Century Oxford." *BSHM Bulletin: Journal of the British Society for the History of Mathematics* 32.2(2017): 149–59.

Gailiunas, Paul. "Mad Weave." *Journal of Mathematics and the Arts* 11.1(2017): 40–58.

Ganzell, Sandy. "Divisibility Tests, Old and New." *The College Mathematics Journal* 48.1(2017): 36–40.

Garelick, Barry. "Why Trendy Math Instruction That Focuses on 'Understanding' Often Cheats Kids." *The Federalist* August 15, 2017.

Gelman, Andrew. "Ethics and Statistics: Honesty and Transparency Are Not Enough." *Chance* 30.1(2017): 37–39.

George, Whitney. "Bringing van Hiele and Piaget Together: A Case for Topology in Early Mathematics Learning." *Journal of Humanistic Mathematics* 7.1(2017).

Golding, Jennie. "Is It Mathematics or Is It School Mathematics?" *Mathematical Gazette* 101.552(2017): 386–400.

Gómez-Collado, María del Carmen, Rafael Rivera Herráez, and Macarena Trujillo Guillén. "Anna Bofill's Use of Mathematics in Her Architecture." *Nexus Network Journal* 19(2017): 239–54.

Gómez-Ferrer, Mercedes, and Yolanda Gil. "Vincenzo Vincenzi, 'Geometer and Engineer to Cardinal Borja.'" *Nuncius* 32.1(2017): 111–45.

Gorroochurn, Prakash. "God Does Not Play Dice: Revisiting Einstein's Rejection of Probability in Quantum Mechanics." *Mathematical Scientist* 42.2(2017): 61–73.

Gould, Peter. "Preventable Errors: From Little Things, Big Things Can Grow." *Australian Mathematics Teacher* 73.4(2017): 29–31.

Goulding, Robert, and Matthias Schemmel. "The Manuscripts of Thomas Harriot (1560–1621)." *BSHM Bulletin: Journal of the British Society for the History of Mathematics* 32.1(2017): 17–19.

Grey, John. "The Modal Equivalence Rules of the Port-Royal Logic." *History and Philosophy of Logic* 38.3(2017): 210–21.

Haffner, Emmylou. "Strategical Use(s) of Arithmetic in Richard Dedekind and Heinrich Weber's *Theorie der algebraischen Funktionen einer Veränderlichen*." *Historia Mathematica* 44.1(2017): 31–69.

Haigh, Thomas. "Colossal Genius: Tutte, Flowers, and a Bad Imitation of Turing." *Communications of the ACM* 60.1(2017): 29–35.

Halback, Volker, and Shuoying Zhang. "Yablo without Gödel." *Analysis* 77(2017): 53–57.

Han, Gang, et al. "The Piecewise Exponential Distribution." *Significance* 14.6(2017): 10–11.

Happersett, Susan. "Artist's Statement." *Journal of Mathematics and the Arts* 11.4(2017): 240–42.

Hardin, Johanna S., and Nicholas J. Horton. "Ensuring That Mathematics Is Relevant in a World of Data Science." *Notices of the American Mathematical Society* 64.9(2017): 986–90.

Harel, Guershon. "The Learning and Teaching of Linear Algebra: Observations and Generalizations." *The Journal of Mathematical Behavior* 46(2017): 69–95.

Hartnett, Kevin. "A Fight To Fix Geometry's Foundations." *Quanta Magazine* Feb. 9, 2017.

Hartnett, Kevin. "Mathematicians Measure Infinities and Find That They Are Equal." *Quanta Magazine* Sept. 12, 2017.

Hartnett, Kevin. "A Path Less Taken to the Peak of the Math World." *Quanta Magazine* June 27, 2017.

Hartnett, Kevin. "Secret Link Found between Pure Math and Physics." *Quanta Magazine* Nov. 1, 2017.

Hartnett, Kevin. "Symmetry, Algebra, and the Monster." *Quanta Magazine* Aug. 17, 2017.

Henle, Jim. "The Same, Only Different." *The Mathematical Intelligencer* 39.2(2017): 60–64.

Henry, Philippe, and Gerhard Wanner. "Johann Bernoulli and the Cycloid: A Theorem for Posterity." *Elemente der Mathematik* 72.4(2017): 137–63.

Herrera, Blas, and Albert Samper. "Generating Infinite Links as Periodic Tilings of the da Vinci–Dürer Knots." *The Mathematical Intelligencer* 39.3(2017): 18–23.

Hitchman, Michael P. "The Story of 8." *Math Horizons* 25.2(2017): 16–17.

Hoerl, Roger W., and Ronald D. Snee. "Statistical Engineering: An Idea Whose Time Has Come?" *The American Statistician* 71.3(2017): 209–19.

Hollings, Christopher D. "'Nobody Could Possibly Misunderstand What a Group Is': A Study in Early Twentieth-Century Group Axiomatics." *Archive for History of Exact Sciences* 71(2017): 409–81.

Hollings, Christopher, Ursula Martin, and Adrian Rice. "The Early Mathematical Education of Ada Lovelace." *BSHM Bulletin: Journal of the British Society for the History of Mathematics* 32.3(2017): 221–34.

Hollings, Christopher, Ursula Martin, and Adrian Rice. "The Lovelace–De Morgan Mathematical Correspondence: A Critical Re-Appraisal." *Historia Mathematica* 44.3(2017): 202–31.

Jeannotte, Doris, and Carolyn Kieran. "A Conceptual Model of Mathematical Reasoning for School Mathematics." *Educational Studies in Mathematics* 96(2017): 1–16.

Jenkins, Alice. "Mathematics." Iin *The Routledge Research Companion to Nineteenth-Century British Literature and Science*, edited by John Holmes and Sharon Luston, New York: Routledge, 2017, 217–34.

Jensen, Jens Højgaard, Martin Niss, and Uffe Thomas Jankvist. "Problem Solving in the Borderland between Mathematics and Physics." *International Journal of Mathematical Education in Science and Technology* 48.1(2017): 1–15.

Jessop, Shunteal. "The Historical Connection of Fourier Analysis to Music." *Mathematics Enthusiast* 14.1–3(2017): 77–99.

Johnson, Ann. "Rational and Empirical Cultures of Prediction." In *Mathematics as a Tool: Tracing New Roles for Mathematics in the Sciences*, edited by Johannes Lenhard and Martin Carrier. Cham, Switzerland: Springer Nature, 2017, 23–36.

Jones, Stephen L. "A Different Perspective of the Teaching Philosophy of R. L. Moore." *International Journal of Mathematical Education in Science and Technology* 48.2(2017): 300–306.

Kainulainen, Mikko, Jake McMullen, and Erno Lehtinen. "Early Developmental Trajectories toward Concepts of Rational Numbers." *Cognition and Instruction* 35.3(2017): 4–19.

Katz, Mikhail, and Luie Polev. "From Pythagoreans and Weierstrassians to True Infinitesimal Calculus." *Journal of Humanistic Mathematics* 7.1(2017).

Kaufholz-Soldat, Eva. "'[...] the First Handsome Mathematical Lady I've Ever Seen!' On the Role of Beauty in Portrayals of Sofia Kovalevskaya." *BSHM Bulletin: Journal of the British Society for the History of Mathematics* 32.3(2017): 198–213.

Kaufmann, Ralph M., and Christopher Yeomans. "Math by Pure Thinking: R First and the Divergence of Measures in Hegel's Philosophy of Mathematics." *European Journal of Philosophy* 25.4(2017): 985–1020.

Kedar, Yael, and Giora Hon. "'Natures' and 'Laws': The Making of the Concept of Law of Nature." *Studies in History and Philosophy of Science, A* 61(2017): 21–31.

Kemp, Gary. "Is Everything a Set? Quine and (Hyper)Pythagoreanism." *The Monist* 100(2017): 155–66.

Kirkpatrick, Keth. "It's Not the Algorithm, It's the Data." *Communications of the ACM* 60.2(2017): 21–23.

Klarreich, Erica. "How To Quantify (and Fight) Gerrymandering." *Quanta Magazine* Apr. 4, 2017.

Knobloch, Eberhard. "Leibniz's Mathematical Handling of Death, Catastrophes, and Insurances." *Matematica, Cultura e Società* 1.3(2016): 259–74.

Knoll, Eva. "Colourwave: Some Variations on a Mathematical Schema." *Journal of Mathematics and the Arts* 11.4(2017): 203–22.

Knudson, Kevin P. "Franz and Georg: Cantor's Mathematics of the Infinite in the Work of Kafka." *Journal of Humanistic Mathematics* 7.1(2017).

Kontorovich, Igor', and Rina Zazkis. "Mathematical Conventions: Revisiting Arbitrary and Necessary." *For the Learning of Mathematics* 37.1(2017): 29–34.

Koss, Lorelei. "Visual Arts, Design, and Differential Equations." *Journal of Mathematics and the Arts* 11.3(2017): 129–58.

Labs, Oliver. "Straight Lines on Models of Curved Surfaces." *The Mathematical Intelligencer* 39.2(2017): 15–26.

Lai, Moon Yee, and Jeffrey P. Wong. "Revisiting Decimal Misconceptions from a New Perspective: The Significance of Whole Number Bias in Chinese Culture." *The Journal of Mathematical Behavior* 46(2017): 96–108.

Lake, Brenden M., et al. "Building Machines That Learn and Think Like People ." *Behavioral and Brain Sciences* 54(2017) [72 pp, including multiple commentaries].

Landy, David, Arthur Charlesworth, and Erin Ottmar. "Categories of Large Numbers in Line Estimation." *Cognitive Science* 41(2017): 326–53.

Langer-Osuna, Jennifer M. "Authority, Identity, and Collaborative Mathematics." *Journal for Research in Mathematics Education* 48.3(2017): 237–47.

Langkjær-Bain, Robert. "Trials of a Statistician." *Significance* 14.4(2017): 14–19.

Lê, François. "Alfred Clebsch's 'Geometrical Clothing' of the Theory of the Quintic Equation." *Archive for History of Exact Sciences* 71(2017): 39–70.

Leibovich, Tali, et al. "From 'Sense of Number' to 'Sense of Magnitude': The Role of Continuous Magnitudes in Numerical Cognition." *Behavioral and Brain Sciences* 54(2017) [62 pp, including multiple commentaries].

Lenowitz, Alan Jay. "The New Astrology." *Aeon Magazine* Apr. 11, 2016.

Lima, Manuel. "Circular Visualizations." *The American Scientist* 105.3(2017): 166–73.

Lindskog, Marcus, Anders Winman, and Leo Poom. "Individual Differences in Nonverbal Number Skills Predict Math Anxiety." *Cognition* 159(2017): 156–62.

Little, John B. "A Mathematician Reads Plutarch: Plato's Criticism of Geometers of His Time." *Journal of Humanistic Mathematics* 7.2(2017).

Liu, Yating, and Mary C. Enderson. "What Does It Mean To Do Something Randomly?" *Mathematics Teacher* 110.5(2017): 364–70.

Lorenat, Jemma. "The Greatest Geometer since the Time of Apollonius." *The Mathematical Intelligencer* 39.3(2017): 46–52.

Lu-Adler, Huaping. "Kant and the Normativity of Logic." *European Journal of Philosophy* 25.2(2017): 207–30.

Luna, Laureano. "Rescuing Poincaré from Richard's Paradox." *History and Philosophy of Logic* 38.1(2017): 57–71.

MacLean, Leonard, and Bill Ziemba. "Sports Statistics: Rating Batters in Baseball." *Wilmott Magazine* 49(2017): 33–34.

Massa-Esteve, M. Rosa. "Mengoli's Mathematical Ideas in Leibniz's Excerpt." *BSHM Bulletin: Journal of the British Society for the History of Mathematics* 32.1(2017): 40–60.

Matt, Andreas Daniel. "IMAGINARY—A How-To Guide for Math Exhibitions." *Notices of the American Mathematical Society* 64.4(2017): 368–74.

Matthews, Robert. "The ASA's *p*-Value Statement, One Year On." *Significance* 14.2(2017): 38–41.

Matthews, Robert J., and Eli Dresner. "Measurement and Computational Skepticism." *Noûs* 51.4(2017): 832–54.

McAndrew, Erica M., Wendy L. Morris, and Francis (Skip) Fennell. "Geometry-Related Children's Literature Improves the Geometry Achievement and Attitudes of Second-Grade Students." *School Science and Mathematics* 117.1–2(2017): 34–51.

McCray, Patrick M. "The Biggest Data of All: Making and Sharing a Digital Universe." *Osiris* 32(2017): 243–63.

Meijer, Erik. "Making Money Using Math." *Communications of the ACM* 60.5(2017): 36–42.

Meyer, Walter. "'Modern' Algebra: Time for Change." *The UMAP Journal* 38.1(2017): 67–71.

Miller, David Marshall. "The Parallelogram Rule from Pseudo-Aristotle to Newton." *Archive for History of Exact Sciences* 71(2017): 157–91.

Miller, Geoffrey, and Samuel Obara. "Finding Meaning in Mathematical Mnemonics." *Australian Mathematics Teacher* 73.3(2017): 13–18.

Miyazaki, Mikio, Taro Fujita, and Keith Jones. "Students' Understanding of the Structure of Deductive Proof." *Educational Studies in Mathematics* 94(2017): 223–39.

Mori, Giuliano. "Mathematical Subtleties and Scientific Knowledge: Francis Bacon and Mathematics, at the Crossing of Two Traditions." *British Journal for the History of Science* 50.1(2017): 10–21.

Morris, Sean. "The Significance of Quine's New Foundations for the Philosophy of Set Theory." *The Monist* 100(2017): 167–79.

Mulcahy, Colm. "The Mathematics Genealogy Project Comes of Age at Twenty-One." *Notices of the American Mathematical Society* 64.5(2017): 466–70.

Müller-Wille, Steffan. "Names and Numbers: 'Data' in Classical Natural History, 1758–1859." *Osiris* 32(2017): 109–28.

Musgrave, Stacy, and Marilyn P. Carlson. "Understanding and Advancing Graduate Teaching Assistants' Mathematical Knowledge for Teaching." *The Journal of Mathematical Behavior* 45(2017): 137–49.

Negro, Antonio, and Carlo Penco. "Kenny's Wrong Formula." *Philosophical Investigations* 40.2(2017): 121–49.

Nelson, Roice, and Henry Segerman "Visualizing Hyperbolic Honeycombs." *Journal of Mathematics and the Arts* 11.1(2017): 4–39.

Netz, Reviel. "Divisions, Big and Small: Comparing Archimedes and Liu Hui." In *Ancient Greece and China Compared*, edited by G.E.R. Lloyd and Jingyi Jenny Zhao, Cambridge, U.K.: Cambridge University Press, 2018, 259–89.

Newman, Mark. "Power-Law Distribution." *Significance* 14.4(2017): 10–11.

Núñez, Rafael E. "Is There Really an Evolved Capacity for Number?" *Trends in Cognitive Sciences* 21.6(2017): 409–24.

O'Donnell, Caroline. "The Deformations of Francesco de Marchi." *Cornell Journal of Architecture* 9(2016): 47–59.

O'Sullivan, Ciaran, et al. "Models of Re-Engaging Adult Learners with Mathematics." *Teaching Mathematics and Its Applications* 36(2016): 81–93.

Oaks, Jeffrey A. "Irrational 'Coefficients' in Renaissance Algebra." *Science in Context* 30.2(2017): 141–72.

Oswald, Nicola M. R. "On a Relation between Modular Functions and Dirichlet Series: Found in the Estate of Adolf Hurwitz." *Archive for History of Exact Sciences* 71(2017): 345–61.

Ottmar, Erin, and David Landy. "Concreteness Fading of Algebraic Instruction: Effects on Learning." *Journal of the Learning Sciences* 26.1(2017): 51–78.

Özgan, Sibel Yasemin, and Mine Özkar. "A Thirteenth-Century Dodecahedron in Central Anatolia: Geometric Patterns and Polyhedral Geometry." *Nexus Network Journal* 19(2017): 455–71.

Padua, Sydney. "Picturing Lovelace, Babbage, and the Analytical Engine: A Cartoonist in Mathematical Biography." *BSHM Bulletin: Journal of the British Society for the History of Mathematics* 32.3(2017): 214–20.

Papadopoulos, Athanase. "Nicolas-Auguste Tissot: A Link between Cartography and Quasiconformal Theory." *Archive for History of Exact Sciences* 71(2017): 319–36.

Parshall, Karen Hunger. "A Plurality of Algebras, 1200–1600: Algebraic Europe from Fibonacci to Clavius." *BSHM Bulletin: Journal of the British Society for the History of Mathematics* 32.1(2017): 2–16.

Passeau, A. C. "Philosophy of the Matrix." *Philosophia Mathematica* 25(2017): 246–67.

Pérez, Joaquín. "A New Golden Age of Minimal Surfaces." *Notices of the American Mathematical Society* 64.4(2017): 347–58.

Persson, Ulf. "Chess and Mathematics." *Svenska Matematikersamfundet Medlemsutskicket* Feb. 2017: 10–12.

Persson, Ulf. "The Mathematics behind Solar Eclipses." *Svenska Matematikersamfundet Medlemsutskicket* Nov. 2017: 11–16.

Perucca, Antonella. "The Chinese Remainder Clock." *The College Mathematics Journal* 48.2(2017): 82–89.

Pierce, David. "On Commensurability and Symmetry." *Journal of Humanistic Mathematics* 7.2(2017).

Pollard, Stephen. "Ethical Guidance from Literature and Mathematics." *The Journal of Speculative Philosophy* 31.4(2017): 517–37.

Prins, Jacomien. "Girolamo Cardano and Julius Caesar Scaliger in Debate about Nature's Musical Secrets." *Journal of the History of Ideas* 78.2(2017): 169–89.

Quinell, Lorna. "Those Muddling M's: Scaffolding Understanding of Averages in Mathematics." *Australian Mathematics Teacher* 73.3(2017): 6–12.

Raddy, Mark. "Why Do They Need To Learn This?" *Australian Mathematics Teacher* 73.3(2017): 4–5.

Radford, Luis, and Wolff-Michael Roth. "Alienation in Mathematics Education: A Problem Considered from Neo-Vygotskian Approaches." *Educational Studies in Mathematics* 96(2017): 367–80.

Raper, Simon. "The Shock of the Mean." *Significance* 14.6(2017): 12–16.

Rasmussen, Chris, et al.. "Four Goals for Instructors Using Inquiry-Based Learning." *Notices of the AMS* 64.11(2017): 1308–11.

Reys, Barbara, and Robert Reys. "Strengthening Doctoral Programs in Mathematics Education: A Continuous Process." *Notices of the American Mathematical Society* 64.4(2017): 386–89.

Rockmore, Don. "What Is Mathematics?" In *What Are the Arts and Sciences? A Guide for the Curious*, edited by Don Rockmore. Dartmouth, NH: Dartmouth College Press, 2017, 220–31.

Rodhe, Staffan. "A Forgotten Booklet by Goldbach Now Rediscovered and Three Versions of Its Contents." *BSHM Bulletin: Journal of the British Society for the History of Mathematics* 32.1(2017): 75–90.

Rosenhouse, Jason. "The Power of a Good Book." *The American Mathematical Monthly* 124.1 (2017): 92–95.

Rosentrater, Ray. "Faking Real Data." *Math Horizons* 25.2(2017): 24–26.

Roth, Wolff-Michael. "Astonishment: A Post-Constructivist Investigation into Mathematics as Passion." *Educational Studies in Mathematics* 95(2017): 97–111.

Rowe, David E. "On Building and Interpreting Models: Four Historical Case Studies." *The Mathematical Intelligencer* 39.2(2017): 6–14.

Saam, Nicole J. "What Is a Computer Simulation?" *Journal for General Philosophy of Science* 48(2017): 293–309.

Sahlén, Martin. "On Probability and Cosmology: Inference beyond Data?" In *The Philosophy of Technology*, edited by Khalil Chamcham et al., New York: Cambridge University Press, 2017, 426–46.

Sak, Ugur, et al. "Creativity in the Domain of Mathematics." In *The Cambridge Handbook of Creativity across Domains*, edited by James C. Kaufmann, Vlad G. Glăveanu, and John Baer. Cambridge, U.K.: Cambridge University Press, 2017, 276–97.

Salsburg, David S. "How Far Is It from the Earth to the Sun?" *Significance* 14.3(2017): 34–37.

Sasanguie, Delphine, et al. "Unpacking Symbolic Number Comparison and Its Relation with Arithmetic in Adults." *Cognition* 165(2017): 26–38.

Schaffer, Simon. "Oriental Metrology and the Politics of Antiquity in Nineteenth-Century Survey Sciences." *Science in Context* 30.2(2017): 173–212.

Schmidt, William H., et al. "The Role of Subject-Matter Content in Teacher Preparation: An International Perspective for Mathematics." *Journal of Curriculum Studies* 49.2(2017): 111–31.

Schubring, Gert. "Searches for the Origins of the Epistemological Concept of Model in Mathematics." *Archive for History of Exact Sciences* 71(2017): 245–78.

Seaton, Katherine A. "Sphericons and D-Forms: A Crocheted Connection." *Journal of Mathematics and the Arts* 11.4(2017): 187–202.

Shapiro, Stewart. "Computing with Numbers and Other Non-Syntactic Things: *De re* Knowledge of Abstract Objects." *Philosophia Mathematica* 25(2017): 268–81.

Shapiro, Stewart. "Mathematics in Philosophy, Philosophy in Mathematics: Three Case Studies." In *Objectivity, Realism, and Proof*, edited by F. Boccuni and A. Sereni, Cham, Switzerland: Springer International Publishing, 2016, 1–9.

Sidman, Jessica, and Audrey St. John. "The Rigidity of Frameworks: Theory and Applications." *Notices of the American Mathematical Society* 64.9(2017): 973–79.

Silver, Daniel S. "The New Language of Mathematics: Is It Possible To Take All Words Out of Mathematical Expressions?" *The American Scientist* 105.6(2017): 364–71.

Simon, Martin A. "Explicating 'Mathematical Concept' and 'Mathematical Conception' as Theoretical Constructs for Mathematics Education Research." *Educational Studies in Mathematics* 94(2017): 117–37.

Singmaster, David. "The Utility of Recreational Mathematics." *UMAP Journal* 37.4(2017): 339–80.

Soberón, Pablo. "Gerrymandering, Sandwiches, and Topology." *Notices of the American Mathematical Society* 64.7(2017): 1010–13.

Sonar, Thomas. ". . . In the Darkest Night That Is . . . Briggs, Blundeville, Wright, and the Misconception of Finding Latitude." *BSHM Bulletin: Journal of the British Society for the History of Mathematics* 32.1(2017): 20–29.

Sørensen, Henrik Kragh. "Studying Appropriations of Past Lives: Using Metabiographical Approaches in the History of Mathematics" *BSHM Bulletin: Journal of the British Society for the History of Mathematics* 32.3(2017): 186–97.

Spallone, Roberta, and Marco Vitali. "Baroque Turin, Between Geometry and Architecture." *The Mathematical Intelligencer* 39.2(2017): 76–84.

Sporn, Howard. "Pythagorean Triples, Complex Numbers, and Perplex Numbers." *The College Mathematics Journal* 48.2(2017): 115–22.

Stark, Benjamin A. "Studying 'Moments' of the Central Limit Theorem." *Mathematics Enthusiast* 14.1–3(2017): 53–75.

Stefánsson, H. Orri. "What Is 'Real' in Probabilism?" *Australian Journal of Philosophy* 95.3(2017): 573–87.

Stewart, Sepideh, and Ralf Schmidt "Accommodation in the Formal World of Mathematical Thinking." *International Journal of Mathematical Education in Science and Technology* 48(2017): S40–S49.

Stonelake, Brian. "Optimally Observing Orbiting Orbs." *Math Horizons* 25.2(2017): 27–29.

Stouraitis, Konstantinos, Despina Potari, and Jeppe Skott. "Contradictions, Dialectical Oppositions and Shifts in Teaching Mathematics." *Educational Studies in Mathematics* 95(2017): 203–17.

Strasser, Bruno J., and Paul N. Edwards. "Big Data Is the Answer: . . . But What Is the Question?" *Osiris* 32(2017): 328–45.

Sutherland, Daniel. "Kant's Conception of Number." *Philosophical Review* 126.2(2017): 147–90.

Tabachnikov, Serge. "Polynomials as Polygons." *The Mathematical Intelligencer* 39.1(2017): 41–43.

Tasic, Vladimir. "Badiou's Logics: Math, Metaphor, and (Almost) Everything." *Journal of Humanistic Mathematics* 7.1(2017).

Tanton, James, and Brianna Donaldson. "The Global Math Project: Uplifting Mathematics for All." *Notices of the American Mathematical Society* 64.7(2017): 712–16.

Tong, Fuhui, and Shifang Tang. "English-Medium Instruction in a Chinese University Math Classroom." In *English-Medium Instruction in Chinese Universities*, edited by Jing Zhao and L. Quentin Dixon, Abingdon, U.K.: Routledge, 128–44.

Tou, Erik. "The Farey Sequence: From Fractions to Fractals." *Math Horizons* 24.3(2017): 8–11.

Tsagbey, Sitsofe, Miguel de Carvalho, and Garritt L. Page. "All Data Are Wrong, but Some Are Useful? Advocating the Need for Data Auditing." *The American Statistician* 71.3(2017): 231–35.

Tunstall, Luke, and Matthew Ferkany. "The Role of Mathematics Education in Promoting Flourishing." *For the Learning of Mathematics* 37.1(2017): 25–28.

Usó-Doménech, José-Luis, et al. "Mathematics, Philosophical and Semantic Considerations on Infinity: Dialectical Vision." *Foundations of Science* 22(2017): 655–74.

van Atten, Mark, and Göran Sundholm. "L.E.J. Brouwer's 'Unreliability of the Logical Principles': A New Translation, with an Introduction." *History and Philosophy of Logic* 38.1 (2017): 24–47.

van Fraassen, Bas C., and Joseph Y. Halpern, "Updating Probability: Tracking Statistics as Criterion." *British Journal for the Philosophy of Science* 68(2017): 725–43.

Vashchyshyn, Ilona, and Egan J. Chernoff. "Sick of Viral Math." *Math Horizons* 25.1(2017): 33–34.

Volkert, Klaus. "On Models for Visualizing Four-Dimensional Figures." *The Mathematical Intelligencer* 39.2(2017): 27–35.

von Oertzen, Christine. "Machineries of Data Power: Manual versus Mechanical Census Compilation in 19th Century Europe." *Osiris* 32(2017): 129–50.

Wainer, Howard. "The Birth of Statistical Graphics and Their European Childhood: On the Historical Development of W.E.B. Du Bois's Graphical Narrative of a People." *Chance* 30.3(2017): 61–67.

Wainer, Howard. "The Grabovsky Curve." *Chance* 30.1(2017): 44–48.

Wallace, Dorothy. "Why I Teach This Subject This Way." *Numeracy* 10.2(2017).

Wardhaugh, Benjamin. "Charles Hutton: 'One of the Greatest Mathematicians in Europe'?" *BSHM Bulletin: Journal of the British Society for the History of Mathematics* 32.1(2017): 91–99.

Wasserman, Nicholas H. "Exploring Flipped Classroom Instruction in Calculus III." *International Journal of Science and Mathematics Education* 15(2017): 545–68.

Waxman, Daniel. "Deflationism, Arithmetic, and the Argument from Conservativeness." *Mind* 126.502(2017): 429–63.

Weaver, George. "König's Infinity Lemma and Beth's Tree Theorem." *History and Philosophy of Logic* 38.1(2017): 48–56.

Weber, Erik, and Joachim Frans. "Is Mathematics a Domain for Philosophers of Explanation?" *Journal for General Philosophy of Science* 48(2017): 125–42.

Weiland, Travis. "Problematizing Statistical Literacy: An Intersection of Critical and Statistical Literacies." *Educational Studies in Mathematics* 96(2017): 33–47.

Weinberg, Andrea, and Laure Beth Sample McMeeking. "Toward Meaningful Interdisciplinary Education: High School Teachers' Views of Mathematics and Science Integration." *School Science and Mathematics* 117.5(2017): 204–13.

Wells, David. "Abstract Games and Mathematics: From Calculation to Analogy." *Svenska Matematikersamfundet Medlemsutskicket* Feb. 2017: 2–9.

Whitty, Robin W. "Some Comments on Multiple Discovery in Mathematics." *Journal of Humanistic Mathematics* 7.1(2017).

Willingham, Daniel T. "Do Manipulatives Help Students Learn?" *American Educator* Fall (2017): 25–31.

Wilson, Robert J. "The Gresham Professors of Geometry." *BSHM Bulletin: Journal of the British Society for the History of Mathematics* 32.1(2017): 125–48.

Woodward, Chris. "Writing a Teaching Letter." *Notices of the American Mathematical Society* 64.10(2017): 1178–79.

Yanofsky, Noson S. "Finding Structure in Science and Mathematics [published as "Chaos Makes the Multiverse Unnecessary]." *Nautilus* June 22, 2017.

Yanofsky, Noson S., and Mark Zelcer. "The Role of Symmetry in Mathematics." *Foundations of Science* 22.3(2017): 495–515.

Yee, Sean P. "Students' and Teachers' Conceptual Metaphors for Mathematical Problem Solving." *School Science and Mathematics* 117.3–4(2017): 146–57.

Yopp, David A., and Jacob L. Ellsworth. "Generalizing and Skepticism: Bringing Research to Practice." *Mathematics Teaching in the Middle School* 22.5(2017): 284–92.

Notable Book Reviews and Review Essays

This list is ordered alphabetically by the reviewer's name. To save space, we use only the main title of the work and we abbreviate the first reviewer's name.

A. Aberdein reviews *The Language of Mathematics* by Mohan Ganesalingam. *Philosophia Mathematica* 26(2017): 143–47.

R. A. Bradley reviews *A Brief History of Numbers* by Leo Corry. *Historia Mathematica* 44(2017): 423–24.

J. R. Brown reviews *An Historical Introduction to the Philosophy of Mathematics*, edited by Russell Marcus and Mark McEvoy. *Philosophia Mathematica* 26(2017): 151–53.

R. Calinger reviews *A Comet of the Enlightenment* [A. J. Lexell] by Johan C.-E. Stén. *The Mathematical Intelligencer* 39.3(2017): 92–93.

R. Chapman reviews *Bridges 2016 Poetry Anthology*, edited by Sarah Glaz. *Journal of Humanistic Mathematics* 7.1(2017): 402–9.

A. I. Dale reviews *Origins of Mathematical Words* by Anthony Lo Bello. *Notices of the American Mathematical Society* 64.1(2017): 47–48.

M. Detlefsen reviews *Why Is There Philosophy of Mathematics at All?* by Ian Hacking. *Philosophia Mathematica* 26(2017): 407–12.

D. DeTurck reviews *The Mathematician's Shiva* by Stuart Rojstaczer. *Notices of the American Mathematical Society* 64.9(2017): 1043–45.

U. Dudley reviews *I, Mathematician*, Vol. 2, edited by Peter Casazza et al. *UMAP Journal* 37.4(2016): 417–18.

J. Evans reviews *Correspondance de Pierre Simon Laplace (1749–1827)* edited by Roger Hahn. *Journal for the History of Astronomy* 48.3(2017): 364–66.

F. Fantini reviews *Origins of Classical Architecture* by Mark Wilson Jones. *Nexus Network Journal* 19(2017): 205–8.

S. C. Fletcher reviews *New Foundations of Physical Geometry* by Tim Mauldin. *Philosophy of Science* 84(2017): 595–603.

M. N. Fried reviews *Mathematicians and Their Gods*, edited by Snezana Lawrence and Mark McCartney. *Mathematical Thinking and Learning* 19.3(2017): 202–7.

S. Gandon reviews *Starry Reckoning* by Emily Rolfe Grosholz. *Philosophia Mathematica* 26(2017): 419–22.

B. Gold reviews *Mathematical Knowledge and the Interplay of Practices* by José Ferreirós. *College Mathematics Journal* 48.3(2017): 226–32.

P. Gorkin reviews *How to Bake Pi* by Eugenia Cheng. *The Mathematical Intelligencer* 39.2(2017): 102–3.

J. V. Grabiner reviews *The Real and the Complex* by Jeremy Gray. *Notices of the American Mathematical Society* 64.11(2017): 1322–26. Yes. Thanks.}

G. Greenfield reviews *Alice's Adventures in Wonderland* [150th anniversary edition] by Lewis Carroll. *Journal of Mathematics and the Arts* 10.1–4(2017).

B. Greer reviews *All Positive Action Starts with Criticism* by Sacha la Bastide-van Gemert. *Educational Studies in Mathematics* 95(2017): 113–22.

E. Grosholz reviews *Realizing Reason* by Danielle Macbeth. *Journal of Humanistic Mathematics* 7.1(2017): 263–74.

G. Hart reviews *Mind-Blowing Modular Origami* by Byriah Loper. *Journal of Mathematics and the Arts* 10.1–4(2017): 180–86.

B. Hayes reviews *The Mathematics of Various Entertaining Subjects*, edited by Jennifer Beineke and Jason Rosenhouse. *Notices of the American Mathematical Society* 64.2(2017): 163–66.

R. Hersh reviews *Elements of Mathematics* by John Stillwell. *American Mathematical Monthly* 124.5(2017): 475–78.

K. F. Hollebrands and S. Okumus review *Tools and Mathematics* by John Monaghan, L. Trouche, and J. M. Borwein. *Journal for Research in Mathematics Education* 48.5(2017): 580–84.

D. Jesseph reviews *Infinitesimal* by Amir Alexander. *The Mathematical Intelligencer* 39.1(2017): 72–73.

K. Jongsma reviews *Sourcebook in the Mathematics of Medieval Europe and North Africa*, edited by Victor J. Katz. *American Mathematical Monthly* 124.7(2017): 667–72.

C. S. Kaplan reviews *Visualizing Mathematics with 3D Printing* by Henry Segerman. *College Mathematics Journal* 48.1(2017): 69–72.

G. Karaali reviews *Mathematics and Art* by Lynn Gamwell. *Journal of Mathematics and the Arts* 10.1–4(2017): 87–92.

V. K. Katz reviews *Elements of Mathematics* by John Stillwell. *Bulletin of the American Mathematical Society* 54.3(2017): 521–28.

A.M.W. Kelly and S. L. Tunstall review (separately) *Numbers and Nerves* by Scott and Paul Slovic. *Numeracy* 10.1(2017).

W. Köberer reviews *Navigation* by J. Bennett. *Annals of Science* 74.4(2017): 331–34.

E. Lamb reviews *From Music to Mathematics* by Gareth E. Roberts. *American Mathematical Monthly* 124.10(2017): 979–82.

D. Lanphier reviews *Summing It Up* by Robert Gross. *American Mathematical Monthly* 124.3 (2017): 282–88.

R. Lawrence reviews *Talking about Numbers* by K. Felka. *History and Philosophy of Logic* 38.4 (2017): 390–94.

J. Lützen reviews *A Brief History of Numbers* by Leo Corry. *The Mathematical Intelligencer* 39.3 (2017): 87–89.

E. Miller et al. review *Large-Scale Data, Big Possibilities,* edited by J. A. Middleton, J. Cai, and S. Hwang. *Journal for Research in Mathematics Education* 48.2(2017): 214–28.

M. Moyon reviews *From Alexandria, through Baghdad,* edited by Nathan Sidoli and Glen Van Brummelen. *Historia Mathematica* 44(2017): 174–78.

Y. Nasifoglu reviews *Architecture and Mathematics from Antiquity to the Future,* edited by Kim Williams and Michael J Ostwald. *BSHM Bulletin: Journal of the British Society for the History of Mathematics* 31.3(2016): 261–65.

D. L. Opitz reviews *Seduced by Logic* by Robyn Arianrhod. *BSHM Bulletin: Journal of the British Society for the History of Mathematics* 31.2(2016): 156–58.

D. J. Pimm and N. Sinclair review *Explaining Beauty in Mathematics* by Ulianov Montano. *The Mathematical Intelligencer* 39.1(2017): 79–81.

S. Pollard reviews *Rethinking Knowledge* by Carlo Cellucci. *Philosophia Mathematica* 26(2017): 413–18.

N. Radakovic and L. Jao review *Creativity and Giftedness,* edited by Roza Leikin and Bharath Sriraman. *Mathematical Thinking and Learning* 19.2(2017): 139–41.

F. Rochberg reviews *A Mathematician's Journeys* [Otto Neugebauer], edited by A. Jones, C. Proust, and J. M. Steele. *Journal for the History of Astronomy* 48.3(2017): 354–57.

J. Roitman reviews *The Thrilling Adventures of Lovelace and Babbage* by Sydney Padua and *Gallery of the Infinite* by Richard Evan Schwartz. *Notices of the American Mathematical Society* 64(2017): 504–7.

D. Rowe reviews *The Oxford Handbook of Generality in Mathematics and the Sciences,* edited by K. Chemla, R. Chorlay, and D. Rabouin. *Isis* (2017): 872–73.

D. Schlimm reviews *Mathematical Knowledge and the Interplay of Practices* by José Ferreirós. *Philosophia Mathematica* 26(2017): 139–43.

R. Siegmund-Schultze reviews *Peter Lax, Mathematician* by Reuben Hersh. *The Mathematical Intelligencer* 39.1(2017): 77–78.

R. Siegmund-Schultze reviews *Recollections of a Jewish Mathematician in Germany* by Abraham A. Fraenkel. *Historia Mathematica* 44(2017): 288–92.

D. S. Silver reviews *Nathaniel Bowditch and the Power of Numbers* by Tamara Plakins Thornton. *The American Scientist* 140.6(2016): 374–78.

E. A. Silver and K. Yankson review *Mathematical Problem Posing,* edited by Florence Mihaela Singer, Nerida F. Ellerton, and Jinfa Cai. *Journal for Research in Mathematics Education* 48.1(2017): 111–15.

I. Volynets reviews *Four Projects* by Stan Allen. *Nexus Network Journal* 19(2017): 547–54.

S. A. Walter reviews *Henri Poincaré* by Jeremy Gray. *Historia Mathematica* 44(2017): 425–35.

S. Williams reviews *The New Math* by Christopher J. Phillips. *The Mathematical Intelligencer* 39.4(2017): 74–76.

P. Zorn reviews *Mathematics and the Real World* by Zvi Artstein. *The Mathematical Intelligencer* 39.4(2017): 72–73.

Notable Teaching Tips

Abd-Elhameed, W. M., and N. A. Zeyada. "A Generalization of Generalized Fibonacci and Generalized Pell Numbers." *International Journal of Mathematical Education* 48.1(2017): 102–7.

Abernethy, Gavin M., and Mark McCartney. "Cannibalism and Chaos in the Classroom." *International Journal of Mathematical Education* 48.1(2017): 117–29.

Azevedo, Douglas, and Michele C. Valentino. "Generalization of the Bernoulli ODE." *International Journal of Mathematical Education* 48.2(2017): 256–60.

Beam, John. "A Powerful Method of Non-Proof." *College Mathematics Journal* 48.1(2017): 52–54.

Bergen, Jeffrey. "Is This the Easiest Proof That nth Roots Are Always Integers or Irrational?" *Mathematics Magazine* 90.3(2017): 225.

Betounes, David, and Mylan Redfer. "The Demise of Trig Substitutions?" *College Mathematics Journal* 48.4(2017): 284–87.

Bradley, David M. "An Infinite Series That Displays the Concavity of the Natural Logarithm." *Mathematics Magazine* 90.5(2017): 353–54.

Brandt, Keith. "Approximations First: A Closer Look at Applications of the Definite Integral." *International Journal of Mathematical Education* 48.1(2017): 94–101.

Capitán, Francisco. "Homographic Pencils for the Ellipse and the Hyperbola." *College Mathematics Journal* 48.2(2017): 134–36.

Carter, Paul, and Yitzchak Solomon. "Relaxing the Integral Test." *College Mathematics Journal* 48.4(2017): 290–91.

Cereceda, José Luis. "Polynomial Interpolation and Sums of Powers of Integers." *International Journal of Mathematical Education* 48.2(2017): 267–77.

Eidolon, Katrina, and Greg Oman. "A Short Proof of the Bolzano–Weierstrass Theorem." *College Mathematics Journal* 48.4(2017): 288–89.

Firozzaman, Firoz, and Fahim Firoz. "Efficient Remainder Rule." *International Journal of Mathematical Education* 48.5(2017): 756–62.

Gkioulekas, Eleftherios. "On the Denesting of Nested Square Roots." *International Journal of Mathematical Education* 48.6(2017): 942–53.

Gordon, Sheldon P. "Visualizing and Understanding l'Hopital's Rule." *International Journal of Mathematical Education* 48.7(2017): 1096–1105.

Gou, Jiangtao, and Fengqing Zhang. "Experience Simpson's Paradox in the Classroom." *American Statistician* 71.1(2017): 61–66.

Hoseana, Jonathan. "Extending the Substitution Method for Integration." *Mathematical Gazette* 101(2017): 538–41.

Jones, Stephen L. "A Different Perspective of the Teaching Philosophy of R. L. Moore." *International Journal of Mathematical Education* 48.2(2017): 300–306.

Kontorovich, Igor'. "Students' Confusions with Reciprocal and Inverse Functions." *International Journal of Mathematical Education* 48.2(2017): 278–84.

Lord, Nick. "An Area Conundrum." *Mathematical Gazette* 101(2017): 325–27.

Lord, Nick. "The Volume of a Cone for Pre-Calculus." *Mathematical Gazette* 101(2017): 534–37.

Lord, Nick. "Using A4-Sized Paper to Illustrate that $\sqrt{2}$ Is Irrational." *Mathematical Gazette* 101(2017): 142–45.

Maltenfort, Michael. "A Function Worth a Second Look." *College Mathematics Journal* 48.1(2017): 55–57.

Martínez, Sol Sáez, Félix Martínez de la Rosa, and Sergio Rojas. "Tornado-Shaped Curves." *International Journal of Mathematical Education* 48.2(2017): 289–300.

Northshield, Sam. "Two Short Proofs of the Infinitude of Primes." *College Mathematics Journal* 48.3(2017): 214–16.

Nystedt, Patrik. "A Proof of the Law of Sines Using the Law of Cosines." *Mathematics Magazine* 90.3(2017): 180–81.

Paparella, Pietro. "A Short and Elementary Proof of the Two-Sidedness of the Matrix Inverse." *College Mathematics Journal* 48.5(2017): 366–67.

Pereira, Leandro da Silva, L. M. Chaves, and D. J. de Souza. "An Intuitive Geometric Approach to the Gauss Markov Theorem." *American Statistician* 71.1(2017): 67–70.

Ramírez, José L., and Gustavo N. Rubiano. "A Generalization of the Spherical Inversion." *International Journal of Mathematical Education* 48.1(2017): 132–49.

Reese, R. Allan. "Graphical Interpretations of Data." *Significance* 14.4–6(2017): 42–43.}

Rock, John A. "A Lecture on Integration by Parts." *Mathematical Scientist* 42.1(2017): 29–37.

Sadhukhan, Arpan. "Partitioning the Natural Numbers to Prove the Infinitude of Primes." *College Mathematics Journal* 48.3(2017): 217–18.

Soares, A., and A. L. dos Santos. "Presenting the Straddle Lemma in an Introductory Real Analysis Course." *International Journal of Mathematical Education* 48.3(2017): 428–34.

Sparavigna, Amelia Carolina, and Mauro Maria Baldi. "Symmetry and the Golden Ratio in the Analysis of a Regular Pentagon." *International Journal of Mathematical Education* 48.2 (2017): 306–16.

Sprows, David. "Unique Factorization and the Fundamental Theorem of Arithmetic." *International Journal of Mathematical Education* 48.1(2017): 130–31.

Stephenson, Paul. "Pappus in Two Dimensions, Mamikon in Three." *Mathematical Gazette* 101(2017): 322–25.

Stephenson, Paul. "Trigonometric Identities from Regular Polygons." *Mathematical Gazette* 101(2017): 327–34.

Tisdell, Christopher C. "Rethinking Pedagogy for Second-Order Differential Equations." *International Journal of Mathematical Education* 48.5(2017): 794–801.

Trenkler, Götz, and Dietrich Trenkler. "Intersection of Three Planes Revisited: An Algebraic Approach." *International Journal of Mathematical Education* 48.2(2017): 285–89.

Wares, Arsalan. "Looking for Pythagoras between the Folds." *International Journal of Mathematical Education* 48.6(2017): 938–41.

Wares, Arsalan, and Iwan Elstak. "Origami, Geometry and Art." *International Journal of Mathematical Education* 48.2(2017): 317–24.

Wolchover, Natalie. "A Long-Sought Proof, Found and Almost Lost." *Quanta Magazine* March 28, 2017.

Zahn, Maurício. "A Deduction of the Golden Spiral Equation via Powers of the Golden Ratio ϕ." *International Journal of Mathematical Education* 48.6(2017): 963–71.

Zhong, Wenti. "On the Eigenvalues of Anticommuting Matrices." *College Mathematics Journal* 48.5(2017): 368–69.

Notable Interviews

This list is ordered alphabetically by the interviewee's name.

S. H. Lui interviews Vladimir Arnold [original in 1997]. *Notices of the AMS* 64.4(2017): 432–38.

Allyn Jackson interviews Michèle Audin. *Notices of the AMS* 64.7(2017): 761–62.

Alexander Diaz-Lopez interviews Arthur Benjamin. *Notices of the AMS* 64.1(2017): 32–33.

Nitis Mukhopadhyay interviews Lynne Billard. *Statistical Science* 32.1(2017): 138–64.

Sylvie Paycha interviews Pierre Cartier. *Mathematical Intelligencer* 39.1(2017): 19–21.

Dalene Stangle interviews Mine Çetinkaya-Rundel. *Chance* 30.2(2017): 54–59.

Davar Khoshnevisan and Edward Waymire interview Mu-Fa Chen. *Notices of the AMS* 64.6(2017): 616–19.

Alexander Diaz-Lopez interviews Thomas Grandine. *Notices of the AMS* 64.2(2017): 130–31.

Hermann Habermann, Courtney Kennedy, and Partha Lahiri interview Robert Groves. *Statistical Science* 32.1(2017): 128–37.

Alexander Diaz-Lopez interviews Ryan Haskett. *Notices of the AMS* 64.3(2017): 239–40.

Diane Timblin interviews Brian Hayes. *American Scientist* 105.5(2017): 312–15.

Alexander Diaz-Lopez interviews Kelsey Houston-Edwards. *Notices of the AMS* 64.8(2017): 870–71.

Leah Hoffmann interviews Subhash Khot. *Communications of the ACM* 60.3(2017): 103–4.

Alexander Diaz-Lopez interviews Evelyn Lamb. *Notices of the AMS* 64.9(2017): 1005–7.

Brian Tarran interviews Julia Lane. *Significance* 14.3(2017): 42–43.

Alexander Diaz-Lopez interviews Reinhard Laubenbacher. *Notices of the AMS* 64.5(2017): 456–60.

Yvonne Dold-Samplonius interviews Bartel Leendert van der Waerden [original in 1993]. *Notices of the AMS* 64.3(2017): 313–20.

Alexander Diaz-Lopez interviews Ciprian Manolescu. *Notices of the AMS* 64.11(2017): 1297–99.

Kate Matsudaira interviews Erik Meijer. *Communications of the ACM* 60.6(2017): 51–60.

Ulf Persson interviews Ives Meyer. *Svenska Matematikersamfundet Medlemsutskicket* May 2017: 8–28.

Toshimitsu Hamasaki and Scott Evans interview Geert Molenberghs. *Chance* 30.1(2017): 16–23.

Ulf Persson interviews David Mumford. *Svenska Matematikersamfundet Medlemsutskicket* Feb. 2017: 20–50.

Amy and David Reimann interview Chris Palmer. *Mathematics Magazine* 90.5(2017): 380–82.

Laure Flapan interviews Lillian Pierce. *Notices of the AMS* 64.10(2017): 1170–71.

Allyn Jackson interviews Kenneth A. Ribet. *Notices of the AMS* 64.3(2017): 229–32.

Alexander Diaz-Lopez interviews Lucas Sabalka. *Notices of the AMS* 64.6(2017): 576–77.

Laure Flapan interviews Karen E. Smith. *Notices of the AMS* 64.7(2017): 718–20.

Joshua Sokol interviews Corina Tarnita. *Quanta Magazine* Dec. 20, 2017.

Martin Raussen and Christian Skau interview Andrew J. Wiles. *Notices of the AMS* 64.3(2017): 198–207.

Notable Lives in Mathematics: Profiles, Memorial Notes, and Obituaries

Charles W. Bachman (1924–2017). *Communications of the ACM* 60.9(2017): 24.

Marcel Berger (1927–2016). *Notices of the AMS* 64.11(2017): 1285–95.

Joan Clarke (1917–1996). *Notices of the AMS* 64.3(2017): 252–55.

Ennio De Giorgi (1928–1996). *Notices of the AMS* 64.9(2017): 1095–96.

Leonard Eugene Dickson (1874–1954). *Notices of the AMS* 64.7(2017): 772–76.

Solomon Feferman (1928–2016). *Bulletin of Symbolic Logic* 23.3(2017): 337–44.

Solomon Feferman (1928–2016). *Notices of the AMS* 64.11(2017): 1254–73.

Steve Fienberg (1942–2016). *Chance* 30.1(2017): 24–25.

Joseph Fourier (1768–1830). *The Mathematical Intelligencer* 39.4(2017): 18–26.

Dmitry Borisovich Gnedenko (1912–1995). *The Mathematical Intelligencer* 39.1(2017): 1–3.

Euphemia Lofton Haynes (1890–1980). *Notices of the AMS* 64.9(2017): 995–1003.

Craig Huneke (1951–). *Notices of the AMS* 64.3(2017): 256–59.

Julius Hurwitz (1847–1916). *The Mathematical Intelligencer* 39.1(2017): 44–49.

Joseph B. Keller (1923–2016). *Notices of the AMS* 64.6(2017): 606–15.

Dusa McDuff (1945–). *Notices of the AMS* 64.8(2017): 892–96.
Maryam Mirzakhani (1977–2017). *The Mathematical Gazette* 101(2017): 545–46.
Aleksander (Olek) Pelczynski (1932–2012). *Notices of the AMS* 64.1(2017): 54–58.
William Playfair (1759–1823). *Significance* 14(2017): 20–23.
Dorothea Rockburne and Max Dehn. *Notices of the AMS* 64.11(2017): 1313–19.
Jean E. Sammet (1928–2017). *Communications of the ACM* 60.7(2017): 22.
Reza Sarhangi (1952–2016). *Journal of Mathematics and the Arts* 10.1–4(2016):1–3.
Jackie Stedall (1950–2014). *BSHM Bulletin: Journal of the British Society for the History of Mathematics* 32.1(2017): 100–111.
Charles P. "Chuck" Thacker (1943–2017). *Communications of the ACM* 60.8(2017): 21.
Richard Lane Tieszen (1951–2017). *Philosophia Mathematica* 25.3(2017): 390–91.
Dana Ulery (1932–). *IEEE Annals of the History of Computing* 39.2(2017): 91–95.
Walter Van Stigt (1927–2015). *Bulletin of Symbolic Logic* 23.1(2017): 122–23.
Magnus J. Wenninge (1919–2017). *Journal of Mathematics and the Arts* 11.4(2017): 243–46.
Andrew Wiles (1953–). *Notices of the AMS* 64.3(2017): 197–227.
Frances Wood (1883–1919). *Significance* 13.5(2017): 35–37.

Notable Journal Issues

This list is ordered alphabetically by the journal name.

"[Research in Mathematics by Undergraduates]." *American Mathematical Monthly* 124.9(2017).
"Mentoring [in Statistics]." *The American Statistician* 71.1(2017).
"Nash." *Bulletin of the American Mathematical Society* 54.2(2017).
"Tate." *Bulletin of the American Mathematical Society* 54.4(2017).
"Modern Slavery." *Chance* 30.3(2017).
"Climate Change." *Chance* 30.4(2017).
"Research-Based Interventions in the Area of Proof." *Educational Studies in Mathematics* 96.2(2017).
"STEM for the Future and the Future of STEM." *International Journal of Science and Mathematics Education* 15.1[suppl](2017).
"Preparing and Implementing Successful Mathematics Coaches and Teacher Leaders." *The Journal of Mathematical Behavior* 46(2017).
"Video as a Catalyst for Mathematics Teachers' Professional Growth." *Journal of Mathematics Teacher Education* 20.5(2017).
"Theoretical Foundations of Engagement in Mathematics: Empirical Studies from the Field." *Mathematics Education Research Journal* 29.2(2017).
"Inferentialism in Mathematics Education." *Mathematics Education Research Journal* 29.4(2017).
"New Directions in Quine Scholarship." *The Monist* 100.2(2017).
"Manifestations of Geometry in Architecture." *Nexus Network Journal* 19.1(2017).
"The Structure, Mathematics and Aesthetics of Shell Structures." *Nexus Network Journal* 19.3(2017).
"Big Data." *NTM Zeitschrift für Geschichte der Wissenschaften, Technik und Medizin* 25.4(2017) [in German].
"Data Histories." *Osiris* 32(2017).
"Abstraction Principles." *Philosophia Mathematica* 25.1(2017).
"Dedekind and the Philosophy of Mathematics." *Philosophia Mathematica* 25.3(2017).
"Teaching Inquiry." *PRIMUS* 27.1–2(2017).

"Perspectives and Experiences in Mentoring Undergraduate Students in Research." *PRIMUS* 27.3—5(2017).

"Inquiry-Based Learning in 1st and 2nd Year Courses." *PRIMUS* 27.7(2017).

"Revitalizing Complex Analysis." *PRIMUS* 27.8—9(2017).

"Breaking Scientific Networks." *Social Studies of Science* 47.3(2016).

"Penalizing Model Component Complexity." *Statistical Science* 32.1(2017).

"Complex Surveys." *Statistical Science* 32.2(2017).

"The Making of Measurement." *Studies in History and Philosophy of Science, A* 65—66(2017).

"Carnap on Logic." *Synthese* 194.1(2017).

"Vagueness and Probability." *Synthese* 194.10(2017).

"Contributions from the SEFI (European Society for Engineering Education) Working Group." *Teaching Mathematics and Its Applications* 36.2(2017).

"Big Data in Higher Education." *Technology, Knowledge and Learning* 22.3(2017).

"Mathematical Creativity and Giftedness in Mathematics Education." *Zentralblatt für Didaktik der Mathematik* 49.1(2017).

"Emotions and Motivation in Mathematics Education." *Zentralblatt für Didaktik der Mathematik* 49.3(2017).

"Applying (Cognitive) Theory-Based Instructional Design Principles in Mathematics Teaching and Learning." *Zentralblatt für Didaktik der Mathematik* 49.4(2017).

"Digital Curricula in Mathematics Education." *Zentralblatt für Didaktik der Mathematik* 49.5(2017).

"Mathematical Tasks and the Student." *Zentralblatt für Didaktik der Mathematik* 49.6(2017).

"Mathematical Creativity and Psychology: Back to the Future?" *Zentralblatt für Didaktik der Mathematik* 49.7(2017).

Acknowledgments

My first thanks go to the authors whose contributions are included in this anthology and to the original publishers of the pieces, for cooperating during the process of republication in this form.

The professional team working on this volume at Princeton University Press is the same as last year: thank you to Vickie Kearn for guidance and advice at all stages of preparing the book; to Lauren Bucca for solving the copyright issues; to Nathan Carr for overseeing the production process; and to Paula Bérard for copyediting the manuscript.

For the first time in twenty-five years (that is, since I arrived in this country), I hold a proper job. Although nominally part-time, the position allows me to teach mathematics abundantly, which I can do aptly and I enjoy greatly. The job also helps me and my family to survive; for a mathematical-speculative mind, surviving in the short term eventually means prospering in the longer term. Thanks to Uday Banerjee and Graham Leuschke for hiring me to teach in the Department of Mathematics at Syracuse University; to Moira McDermott and Jeff Meyer for assigning me classes scheduled conveniently; and to Julie O'Connor, Kim Ann Canino, and Sandra Ware for their prompt assistance every time I need administrative help.

My family has been through extreme difficulties for more than a decade, but things are getting better. Lately, I found the time to work on this book only because of the domestic help we received from my parents-in-law, Yanguang Li and Xianhua Lu. Thank you to all at home: my wife Fangfang, my daughter Ioana, and my sons Leo and Ray—especially to Ray, our newcomer in the world, arrived shortly after I wrote my acknowledgments to the previous volume.

Credits